I0180846

Dammed Divinities

The Water Powers at Bujagali Falls, Uganda

Terje Oestigaard

NORDISKA AFRIKAINSTITUTET, UPPSALA 2015

INDEXING TERMS:
Uganda
Nile River
Dams
Water power
Conflicts
Traditional religion
Cultural heritage
Belief
Rituals

ISSN 0280-2171
ISBN 978-91-7106-768-5
© The author, The Nordic Africa Institute.
Cover photo: Traditional healer Jaja Bujagali. Photo taken by the author.
Production: Byrå4
Print on demand. Lightning Source UK Ltd.

Contents

List of figures

Acknowledgements

I would first and foremost like to thank all the informants who generously gave of their time and answered my questions, in particular Jaja Bujagali, Jaja Itanda, Jaja Kagulu and Jaja Kiyira. There are many others I interviewed in the course of my fieldwork, and special thanks are due to Richard Gonza and Waguma Yasin Ntembe for putting me in touch with key informants. Simon provided invaluable help as a translator and Moses, my assistant throughout, deserves special thanks.

A research project does not take place only in the field. The Uganda National Council for Science and Technology provided me with a research permit (SS 3202). The Nile Basin Research Programme (NBRP) also assisted me, and I particularly thank Prof. Edward Kirumira, Margaret Kyakuwa and the director of NBRP, Dr Tore Saetersdal. Not least my thanks are due to Prof. Terje Tvedt. The Nordic Africa Institute, Uppsala, my workplace, has also supported this project and provided a stimulating research environment. I would like to thank all the staff and in particular Prof. Kjell Havnevik. Lastly, I thank Pernilla Bäckström for her invaluable help with library loans and Peter Colenbrander for commenting upon the language.

Especially when studying the controversies surrounding dam building, there is literally a flood of opinion on the web. Not all of this information has been relevant, but some of the most interesting discussions and data are to be found online and not in published papers. Regarding references, endnotes have been used to make the text easier to read. References to internet blogs, newspaper articles online and unpublished reports are included only in the endnotes, whereas published works are also included in a separate bibliography. Given the enormous volume of data dams generate, I have tried to limit the sources to what I perceive to be the most relevant. As always, I am solely responsible for the content and the interpretations. Unless otherwise stated, I have taken the photos.

Terje Oestigaard
Uppsala, 26 February 2015

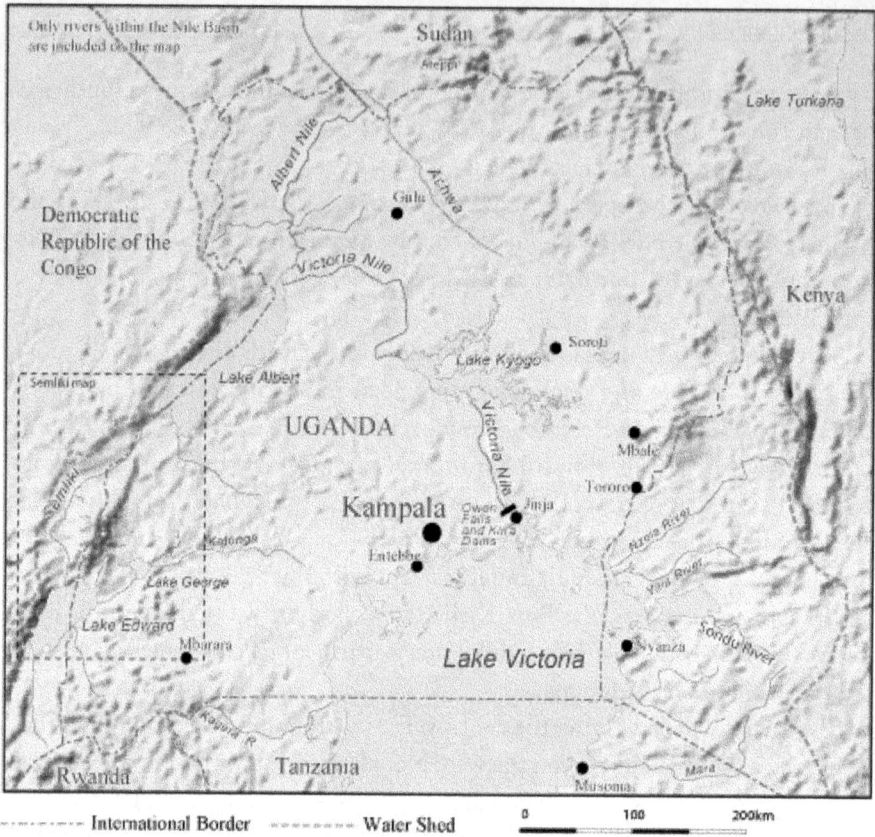

Figure 1. Map of Uganda and Lake Victoria. Nile Basin Research Programme, University of Bergen, Norway.

The Bujagali Hydropower Project in Uganda has been one of the most controversial dam projects in modern times. Located some eight kilometres north of Jinja and the outlet of Lake Victoria or the historic source of the White Nile (Fig. 1), it was Uganda's second large dam when it was inaugurated in 2012. More dams are planned along the Nile, including at Itanda or Kalagala, the next waterfall north of Bujagali, some 30 kilometres from Jinja. Apart from the general criticism of dams, the Bujagali project has been unique in the sense that the controversies surrounding it relate to the major water spirit in the culture of the Busoga Kingdom, whose home was to be flooded. In fact, the dam was postponed for years because of this spirit and the fierce disputes between two healers claiming to be the water spirit's representatives.

Water powers take many forms and in the Bujagali Falls one may see how technology and cosmology are interlinked in local and global discourse. The analysis in this publication takes three key issues as its central focus.

First, water is the basis for hydropower, and Uganda's industrialisation depends on sufficient energy. The Bujagali Dam is crucial in increasing the country's energy supply, as are the other dams along the Nile in Uganda that are being planned or constructed.

Second, the power of water is manifest in the waterfalls themselves, which are testimony to the river spirits living there. Throughout Uganda, water and in particular waterfalls are part of the cosmology of traditional, indigenous religions, which take many forms. In the Bujagali Falls, there are innumerable spirits, the most powerful being the Budhagaali spirit, which is embodied in one particular healer, Jaja Bujagali, who is perceived to be an 'archbishop' in the traditional religion.

Third, water has the power to generate powerful discourses locally and internationally. The construction of large dams causes intense and heated debate between water developers and anti-dam campaigners. Although proponents and opponents share the same overall goal, namely the betterment of people and society, they differ as regards the best development practices, with or without foreign aid or capital.

This analysis contextualises the Bujagali project both historically and globally. Why has it been seen as so controversial? What was the ritual drama behind the scenes when three major appeasement ceremonies were conducted to secure the Budhagaali spirit's blessing of the dam? How did the dam affect the main river spirit and indigenous religion, and is it possible to drown a river spirit in its own element – water? All these questions relate to modernity and development discourses, but with regard to Bujagali, there was one additional factor: How do you negotiate with a spirit?

The analysis starts with an historical context. Historically, the British explorer John Hanning Speke discovered the source of the White Nile in Uganda on 28 July 1862. Although his discovery was challenged back in England and is so even today, since it was of the outlet of Lake Victoria and not the remotest source, it was an historic event, and, by linking the Great Lakes with the Mediterranean, was to shape the development of a large part of Africa. Importantly, it has to be seen in relation to British colonisation of East Africa. More specifically, the first ideas of damming the Nile in Uganda were put forward as early as the second half of the nineteenth century. Then the analysis turns to the construction of the Owen Falls Dam and the Bujagali Dam, including facts and figures. The Bujagali Dam is set within the context of the World Commission on Dams and its report of 2000. The differing positions of dam developers and anti-dam campaigners will be discussed, as will China's role as a new player in the global dam business. There follow a thorough description and analysis of what happened specifically with regard to the Budhagaali water spirit residing in the waterfalls. The construction team aimed to get local acceptance of the dam by arranging grandiose appeasement ceremonies for the spirit, and no fewer than three such ceremonies took place over the next 10 years. However, since spirits sign no agreements and certificates vouching that they are satisfied, such evidence has to be secured by the healers embodying the spirits. But can a healer sign an agreement on behalf of a god? And who is the spirit's proper medium? The analysis ends by addressing the religious consequences of the dam and what happened to the spirit, and also other potential religious challenges when new dams are built along the Nile in Uganda.

The source of the White Nile

The searches for the source of the White Nile in the nineteenth century were altogether different from the searches that uncovered the source of the Blue Nile in terms of implications and ramifications for world history, and the consequences even today. While the source of the Blue Nile in Ethiopia was probably first documented by Portuguese missionaries in 1615[1] and its discovery claimed by the Scottish explorer James Bruce in 1770, it was clear that the Nile quest was not yet complete. The most important search was yet to be undertaken. Dr Charles Beke noted in 1847 that 'the position of the source of that celebrated river remains as unknown as it was in the earliest ages … its head is still enveloped in the clouds of mystery which have in all ages concealed it from our sight.'[2] In 1848 and 1849, however, there came reports from German missionaries in East Africa of sightings of snowcapped mountains close to the Equator. Although these reports were immediately dismissed back in London – such mountains at the Equator were an impossibility – a nagging question remained: could these be the legendary Mountains of the Moon described by Ptolemy 1,700 years earlier? If so, were the mountains feeding the large African inland seas which nobody in the West had seen, but which were depicted on Ptolemy's map? By finding the lakes, one would find the source of the Nile, it was believed.

The quest for the source of the White Nile involved larger-than-life characters such as David Livingstone, Richard Francis Burton, John Hanning Speke, Samuel White Baker and Henry Morton Stanley. Speke has achieved eternal fame as the acknowledged discoverer of the White Nile's source. Although his discovery did not go unchallenged back in Britain, this was a decisive moment in the history of Eastern Africa. As Sir Roderick Murchison, the President of the Royal Geographical Society in London, claimed in 1863, 'the grand achievement of Speke and Grant, who, by traversing a region never previously explored by civilised man, have solved the problem of ages; and have determined that the great fresh-water lake Victoria Nyanza, whose southern watershed extends to three degrees south of the equator, is the reservoir from which the sacred Bahr-el-Abiad, or White Nile, mainly descends to Gondokoro, and thence by Khartoum into Egypt.'[3]

Approaching the source of the Nile, or the outlet of Lake Victoria, on 21 July 1862, Speke writes:

> Here at last I stood on the brink of the Nile; most beautiful was the scene, nothing could surpass it! It was the very perfection of the kind of effect aimed at in a highly-kept park; with a magnificent stream from 600 to 700 yards wide, dotted with islets and rocks, the former occupied by fishermen's huts, the latter by sterns and crocodiles basking in the sun, – flowing between fine grassy

Figure 2. John Hanning Speke and the source of the White Nile, 28 July 1862. Picture at Speke Hotel, Kampala.

banks, with rich trees and plantains in the background, where herds of nsunnŭ and hartebeest could be seen grazing, while the hippopotami were snorting in the water, and florikan and guinea-fowl rising at our feet.[4]

When coming to the outlet of Lake Victoria on 28 July 1862, the search for the source was over (Fig. 2):

> We were well rewarded; for the 'stones,' as the Waganda call the falls, was by far the most interesting sight I had seen in Africa … Though beautiful, the scene was not exactly what I expected; for the broad surface of the lake was shut out of view by a spur of hill, and the falls, about 12 feet deep, and 400 to 500 feet broad, were broken by rocks. Still it was a sight that attracted one to it for hours – the roar of the waters, the thousands of passenger-fish, leaping at the falls with all their might, the Wasoga and Waganda fishermen coming out in boats and taking post on all the rocks with rod and hook, hippopotami and crocodiles lying sleepily on the water, the ferry at work above the falls, and cattle driven down to drink at the margin of the lake, – made, in all, with the pretty nature of the country – small hills, grassy-topped, with trees in the folds, and gardens on the lower slopes – as interesting a picture as one could wish to see.[5]

He continues: 'I now christened the "stones" Ripon falls, after the nobleman who presided over the Royal Geographical Society when my expedition was got up; and the arm of water from which the Nile issued, Napoleon Channel, in token of respect to the French Geographical Society, for the honor they had done me, just before leaving England, in presenting me with their gold medal for the discovery of the Victoria N'yanza.'[6] The source of the White Nile that Speke documented in 1862 is located in Jinja, Uganda, at the outlet of Lake Victoria, where the White Nile starts flowing northwards. As Moorehead notes, 'It could still be argued, of course, that the ultimate source of the headwaters of the main stream that feeds Lake Victoria – [is] Kagera … For ordinary purposes it would seem most sensible to accept the side of the Ripon Falls as the source, since it is only from there that the mighty river confines itself to a definite course …'[7]

The source of the White Nile or the outlet of Lake Victoria has of course always been well known to the people living there, and the closer their proximity to the outlet and the falls, the greater the importance of the place in their local life-worlds. But the importance of the outlet of such a mighty lake as Lake Victoria depends largely on where one lives along the lake or its tributaries. And while the meaning and function of the source differed greatly within and among the former Ugandan kingdoms, the differences in outlook and belief were even more marked between those kingdoms and the Europeans. In particular, the British in the latter half of the nineteenth century had a radically different view of the source of White Nile and the outlet of Lake Victoria than the local, indigenous population.

Sources are not just any body of water, but precisely because they are sources of a river – and in the case of the Nile the longest river in the world (although nobody knew that in 1862) – they are literally the source of what comes later – the river proper. In many respects, the search for and knowledge of the sources are more important for downstream users and countries than for those living nearby, who may not know that the river flows downstream for thousands of kilometres to other people and lands who depend on it in unimaginably different ways, such as Egypt, located in a desert, and drawing 97–98 per cent of its water from the Nile. And from a downstream perspective, from the Egyptian dynasties to the British imperialists, the sources take on very different meanings and importance. Not only were the origins of the mighty river a source of mythology and uncertainty – whence did the water come, why and when, including religious perceptions and beliefs – but knowing the answers to these questions was also a source of understanding and of organising downstream societies and securing the livelihood of millions, whether faith was placed in science (dams and irrigation) or in gods.

From this perspective, Speke becomes the discover of the source of the White Nile, not because it had never been 'discovered' before by Arab traders or more importantly by indigenous people living nearby, but because locating the source only has meaning as a 'discovery' in a particular context and knowledge system. And within this knowledge system, it was acknowledged that Lake Victoria was connected to the Mediterranean. Since antiquity, it has been the Mediterranean that linked the quest for the sources to Europe, culminating with the British imperial explorers. Still, Speke's discovery was contested back in England at the time, and finally confirmed by Stanley in 1875 when he proved that there are no other outlets from Lake Victoria than the one Speke documented in 1862. It is in the context of river systems linking Lake Victoria to the Mediterranean that Speke's claim of discovering the Nile's source is to be understood. And it is only in this context that Speke claimed to have made a discovery.

Explorers and colonials

Apart from personal ambition and possible eternal fame for the discoverer of the Nile's sources, there were much more to this obsession, not least from an imperial perspective. British interest in the source of the Nile was unrelated to local cultural and cosmological perceptions of the water, but was related in part to the scientific quest to solve one of the great unanswered geographical questions of all time, as well as to the prospect of future commerce. After Burton and Speke's expedition (1857–59), when Speke made a detour and for the first time saw the southern reaches of Lake Victoria, there remained as many unanswered questions as before, although new knowledge had been gained. One such question was whether the monsoon that gave rise to the annual flooding of the Blue Nile reached as far south as Speke claimed. 'The solution of this question alone is worthy of the labours of another expedition,' Colonel Sykes commented. 'The work is only half accomplished, and the reputation of our country demands that it should be completed. My own opinion is, that independent of any commercial advantages or sordid considerations, the Society ought, for the simple investigation and verification of physical truths, to use its best endeavours to induce the Government to send out a second expedition. For the good name of England; let us have the doubt removed. We have an inkling of the truth; let us have the whole truth.' He was specifically preoccupied with the Mountains of the Moon and whether the western source of the Nile issued from lakes below the mountains.[8]

Perhaps more importantly, the quest was also about colonisation and imperialism. 'The progress of civilization depends upon geographical position. The surface of the earth presents certain facilities and obstacles to general access; those points that are easily attainable must always enjoy a superior civilization to those that are remote from association with the world,' Baker said.[9] The role of commerce in the civilisation process was eloquently elaborated by John Stuart Mill:

> The opening of a foreign trade, by making them acquainted with new objects
> ... sometimes makes a sort of industrial revolution in a country whose resources
> were previously underdeveloped for want of energy and ambition in the people:

inducing those who were satisfied with scant comforts and little work, to work harder for the gratification of their new tastes, and even to save and accumulate capital, for the still more complete satisfaction of those tastes at a future time ... [Commerce] is one of the primary sources of progress ... rapidly rendering war obsolete, by strengthening and multiplying the personal interests which are in opposition to it.[10]

Or, in the words of Baker: 'The philanthropist and the missionary will expend their noble energies in vain in struggling against the obtuseness of savage hordes, until the first step towards their gradual enlightenment shall have been made by commerce. The savage must learn to *want;* he must learn to be ambitious; and to covet more than the mere animal necessities of food and drink.'[11]

These explorers were also militant geographers,[12] or, again in Baker's words: 'The primary object of geographical exploration is the opening to general intercourse of such portions of the earth as may become serviceable to the human race. *The explorer is the precursor of the colonist*; and the colonist is the human instrument by which the great work must be constructed – that greatest and most difficult of all undertakings – the civilization of the world.'[13]

In 1909, William Gartin, the British mastermind of hydrological engineering along the Nile, pointed out that Speke's discovery of the source of the Nile was 'an event which has produced such far-reaching effect – politically as well as geographically.'[14] According to Gartin, the first phase of the story commences with Burton and Speke's expedition of 1856 and ends with the conquest of the Sudan by Lord Kitchener in 1898. During this period, the Nile sources were discovered, the river basin was explored and the partition of Africa among the European powers took place. In the ten years after 1898, no major geographical discoveries were made, but much useful work had been accomplished, including the completion of the first Aswan High Dam in Egypt.[15]

With a focus on the Nile, British imperialism in this part of Africa can be understood in different terms. Conventionally, it has been argued that the British marched upstream because of the 'frontiers of fear.' By contrast, Terje Tvedt argues it was the limits of irrigation water in Egypt on one hand and the abundance of Nile waters to be controlled for the benefit of Egypt and cotton production on the other that underlay imperial strategy.[16] As Tvedt asks, why were the British more interested in the modest White Nile than in the mighty Blue Nile, which provides much more water, and why did they argue that it was just a matter of time before they had to occupy Sudan?[17] And why did they colonise Uganda?

In his *The River Nile in the Age of the British. Political Ecology and the Quest for Economic Power* (2004), Tvedt analysed in detail how British imperial policies regarding the Nile had the single overall aim of securing British interests.

Britain took direct control of Egypt in 1882. Small amounts of cotton produced in the delta had been sold on the growing world market from the 1820s, but from the 1860s cotton made up 80 per cent of Egypt's exports. British industry had huge economic interests in Egypt. The Lancashire textile industry aimed to reduce its dependence on American cotton and increase the supply of cheaper Egyptian cotton. British banks had a strong interest in the Egyptian economy. By 1882 Egypt's foreign debt had increased to £100 million, with an annual debt amounting to £5 million, of which a large part went to Britain. 'Nile water awareness' in London was so great that *The Times* reported regularly on the Nile's discharges.[18] Wealth depended upon water. Egyptian Prime Minister Nubar Pasha (1884–88 and 1894–95) summarised Egypt's situation in a famous one-liner: 'The Egyptian question is the irrigation question,'[19] a sentiment that echoes Herodotus's much earlier observation that 'Egypt is the gift of the Nile.'

To increase productivity and profit, more Nile water was needed at the 'timely season' or the summer season in Egypt. This is, however, before the arrival of annual Blue Nile floodwaters. Consequently, cotton production was dependent on the White Nile. The overriding challenge was how to secure enough water for cotton production and at the same time control potentially devastating floods. Thus, by the end of the 19th century, 'increased and improved water control was destined to top the agenda of any administration in Egypt.'[20] The British were the first to truly see this connection, and their colonisation of this part of Africa cannot be properly understood without a water perspective focused on the Nile river system.

The British administration under Lord Cromer understood this from the outset and British policy was structured around the Nile. He wrote: 'When, eventually, the waters of the Nile, from the Lakes to the sea, are brought fully under control, it will be possible to boast that Man, in this case the Englishman, has turned the gifts of Nature to the best possible advantage.'[21] Arguing in favour of the construction of the Aswan Dam, Cromer noted that it was of 'utmost importance' because 'the prosperity of Egypt depends wholly on the Nile.' Scott-Moncrieff summed up the 'Nile vision' of the water planners in 1895: 'Is it not evident, then, that the Nile from Victoria Nyanza to the Mediterranean should be under one rule?'[22] Frederick Lugard, later Lord Lugard, who was the British administrator of Uganda, wrote in the early 1890s: 'Egypt is indebted for her summer supply of water to the Victoria lake, and a dam built across the river at its outlet from the lake would deprive Egypt of this.' He continued that the 'occupation of so distant a point as Uganda would be a fair and just claim to render valid our influence in the Nile basin and beyond.'[23] Controlling Uganda and the waters flowing from the source was seen as a geopolitical tool for controlling Egypt and of vital importance to the British. 'The British decision to occupy the Upper Nile should most fruitfully be seen as an example of a far-

Figure 3. Winston Churchill standing on an observation ladder at Hippo Camp in Uganda. Source: Churchill 1909, facing p. 182.

sighted imperial expansionist policy, driven by a complex mixture of economic and political considerations framed by the Nile's geographical and hydrological characteristics,' Tvedt argues. The British had 'become rulers of a truly hydraulic society, where stability and wealth depended upon the water of the Nile.'[24]

Churchill on the Nile

Of all the great statesmen who have been on the Nile, it is perhaps Winston Churchill who most eloquently pointed out its role and importance. The young Churchill visited Uganda protectorate in 1907, using his preferred mode of transport, the bicycle. The colonial mindset required that a guest was provided with a successful lion hunt. On such occasions, the colonialist felt a deep obsession about and responsibility for tracking a lion and killing it (Fig. 3); 'How to find, and having found, to kill, a lion is the unvarying theme of conversation; and every place and journey is judged by a simple standard – "lions or no lions",'[25] Churchill wrote. Pleasure aside, business came first. In Uganda, Churchill asked: 'What is to be their [the Africans'] part in shaping the future of their country? It is, after all, *their* Africa. What are they going to do for it, and what is it going to do for them?'[26]

The current state of affairs was described in a less flattering way: 'And to compare the life and lot of the African aboriginal – secure in his abyss of contended degradation, rich in that he lacks everything and wants nothing – with the long nightmare of worry of privation, of dirt and gloom and squalor, lit only by gleams of torturing knowledge of tantalizing hope, which constitutes the lives of so many poor people in England and Scotland, is to feel the ground tremble under foot.'[27] And the remedy prescribed by Churchill? 'Indeed, it would be hard to find a country where the conditions were more favourable than in Uganda to a practical experiment in State Socialism.'[28]

When Churchill's journey came to an end, he asked rhetorically what message he could take back: 'It can be stated in three words. Concentrate upon Uganda!' Apart from Egypt, 'there is no region which offers prospects to compare in hopefulness with those of the Protectorate of Uganda.' And he adds: 'Nowhere else in Africa will a little money go so far. Nowhere else will the results be more brilliant, more substantial or more rapidly realized.'[29] The legacy of Churchill's optimism about Uganda still lingers, even in tourist advertisements, which describe Uganda as the 'The Pearl of Africa,' a phrase generally attributed to Churchill. But was this one-liner ever uttered by Churchill? He said: 'Yet it is not possible to descend the Nile continuously from its source at Ripon Falls without realizing that the best lies behind one. Uganda is the pearl.'[30] He did not write that Uganda is 'The Pearl of Africa.'

In fact, this phrase seems to be owing to Stanley. Fredrick Lugard wrote in 1893 that 'Stanley was even louder in his praises of Uganda than Speke had been, and described it as the "Pearl of Africa." In consequence of his appeal on behalf of the people … missionaries … arrived in June 1877.'[31] And Stanley writes in his *Africa: Its Partition and Its Future* (1898), 'Uganda, the pearl of Africa, discovered by Mr. Stanley, snatched by Captain Lugard from the hands of the French, and now in the throes of a mutiny, is the cockpit of Central Africa … It is the land of romance and of the unexpected. It commands the northern shores of the Victoria Nyanza and the head waters of the Nile.'[32]

Developing Jinja for hydropower was part of the colonial plan for Uganda. 'Jinja is destined to become a very important place in the future economy of Central Africa. Situated at the point where the Nile flows out of the Great Lake, it is at once the easiest line of water communication with Lake Albert and the Soudan … also where great waterpower is available,' Churchill writes, 'there is power enough to gin all the cotton and saw all the wood in Uganda, and it is here that one of the principal emporia of tropical produce will certainly be created.'[33] He wrote in 1908 that the Ripon Falls were worth visiting just for the beauty of the place (Fig. 4). But, more importantly: 'It is possible that nowhere else in the world could so enormous a mass of water be held up by so little masonry. Two or three short dams from island to island across the falls would enable, at an inconceivably small cost, the whole level of the Victoria Nyanza.'[34] Churchill continues:

Figure 4. The 'stones' or the Ripon Falls in the 1940s. Source: Wikimedia Commons, Dudley Essex.

As one watches the surging waters of the Ripon Falls and endeavours to compute the mighty energies now running to waste, but all within the reach of modern science, the problem of Uganda rises in a new form on the mind. All this water-power belongs to the State. Ought it ever to be surrendered to private persons? How long, on the other hand, is a Government, if not prepared to act itself, entitled to bar the way to others ... This is not to advocate the arbitrary exclusion of private capital and enterprise from Uganda. Careful directed and narrowly controlled opportunities for their activities will no doubt occur. But the natural resources of the country should, as far as possible, be developed by the Government itself.[35]

Churchill could not have been truer. The Bujagali Dam is the largest private sector investment in Uganda, although it was developed in cooperation with the government of Uganda. It was not Churchill, however, who first proposed the idea of dam building in Uganda. The first suggestion of damming Lake Victoria came from Stanley. He wrote in 1878, as he stood at Ripon Falls, where there were four separate waterfalls:

While standing on the Usoga side of the falls, photographing and taking a ground plan of them, I thought what an immense difference might be effected in the level of the Victoria Falls if Mtsea were assisted by engineers. He might by a couple of charges of dynamite, and the destruction of the two islands between which the surplus water rushes into the Victoria Nile, reduce the lake by 12 feet; or he might, by employment of the vast labour at his disposal, dam up the gaps which nature has created, and soon extend the lake thousands of square miles! A dam 50 feet high would transform most of the Luwamberri plain into a broad lake.[36]

But the splendid beauty of the Nile was also unquestionable, 'Looking down the broad stream … one could scarcely wish to behold a more beautiful or more inspiring scene than that which the full-born Nile has created.'[37] However, business came before beauty, and from an interesting note by the president of the Royal Geographical Society made after Stanley's presentation, it seems even King Mutesa was not averse to damming the Nile. 'M'tesa had a notion that his importance and independence might be secured by entirely barricading the Nile, by erecting a vast dam across the Ripon Falls, so as to keep the Victoria Nyanza to himself. Whether that was likely to produce a result favourable to east Africa, he did not know,' the president said. He continued: 'when companies were formed for the purpose of flooding the Sahara, perhaps English capitalists might do something to help M'tesa to flood the lowlands to the east of Speke Gulf, and to bring the Indian Ocean nearer to the Victoria Nyanza.'[38]

The Nile is Wealth

In both the Lusoga and Luganda languages Jinja means 'rock,' and this 'rock' is often identified as the rocks in the river where the boats embark. However, 'rocks' or 'stones' is 'Mayinja,' and Jinja is singular, directing attention to a particular stone, and not any rocks in the Ripon Falls or near the source. Jinja derives from one ancestor; Jinja's great-great-grandfather, whose original name was Idinda. In the courtyard of a house located atop the eastern slopes overlooking the Nile, behind a fenced wall, there is a large stone. When struck, it sounds like a real drum. In the past, when people were going to canoe across the Nile, they would start from this place. Crossing the river just above the roaring rapids was a dangerous enterprise because of the force of the river. To make a safe crossing, travellers needed spiritual protection from the Idinda ancestor. Today, a healer from a distant part embodies this spirit and has his shrine next to the sacred stone. Until recently, there was a sacred tree in which the spirit resided, but it has been cut down.

Still, Jinja has been seen as an 'alien' city and mainly the creation of Indians and Europeans. Although today Jinja is an industrial centre, the overwhelming majority of Basoga have been rural. In 1950, of some 20,000 people living in Jinja, almost 6,000 were non-African and only about 30 per cent were born in the district.[39] The current population of Jinja municipality is about 80,000, but swells each day to an estimated 2–300,000 due to labour commuting.

In the Busoga Kingdom, there is a Lusoga proverb that 'The Nile is Wealth.'[40] The proverb captures the essence of water. The modern city of Jinja has the slogan 'Kiyira gives richness,' meaning that the river gives everything. Jinja's municipal armorial bearings contain the same motto, in Luganda *Kiyira bwe bugagga*. In 1963, the official presentation and interpretation of the emblem was:

> The hippopotamus is representative of the fauna of the district. The rock, or stone, is a reference to the name Jinja which in Luganda means the stone and is thought to have historical significance. The wavy bar ... denotes the river Nile at the source of which Jinja stands. The cotton plants refer to one of the principal crops of Busoga District in which Jinja is situated and the cogwheel and flash of lightning allude to industrial development and the Owen Falls Hydro-electric scheme on the Nile, which is the source of energy for the industry and the country in general. The group of a shield, spears, drums and an antelope's head is a representation of the badge used by the Busoga African Local Government.[41]

As this extract highlights, the wealth of Jinja is intrinsically linked to hydropower, which generates both electricity and money. Between 1945 and 1990, more than 1,000 large dams – of at least 15 metres in height or with a reservoir capacity of 3 million cubic metres or more – were built in Africa for irrigation and electricity. These dams supply 22 per cent of Africa's total electricity, but countries like Mozambique, Rwanda, Tanzania, Uganda and Zambia obtain more than 80 per cent of their power from dams.[42]

Although the Owen Falls Dam was inaugurated in 1954, this was only the first of the dams at the source and along the Nile. In 2000, the first turbines were installed in a second dam, Owen Falls Extension, just below the Owen Falls Dam. During the official commissioning ceremony, President Museveni renamed Owen Falls Dam Nalubaale Hydro Power Station and Owen Falls Extension Kiira Hydro Power Station.[43]

The Bujagali Dam was inaugurated in 2012. Itanda or Kalagala is the next waterfall north of Bujagali, some 30 kilometres from Jinja.

The Budhagaali spirit is part of a wider water cosmology. Among Busoga, all three falls have special religious significance. In each of the falls resides a river spirit or god. In fact, there are innumerable spirits, but there are three main spirits. The three main river spirits incarnate themselves in human form as a traditional healer or diviner. The Bujagali spirit is the supreme god and is seen as the father or elder brother of the other two. This is due to the force of the falls, naturally and hence spiritually. Thus, the force and sound of the rapids testifies to the presence and power of the spirits: they can remove any malignance or misfortune, or be malevolent if devotees are disobedient.

Because the spirit is embodied in them, traditional healers can transfer and mediate the spirit's power. Believers or clients come to the healer with requests for wealth and success or health and prosperity for their children. The healer as medium asks the spirit or god to respond to these requests, and if there are successful physical or spiritual outcomes, the believer will fulfil his or her promise and obligation to the god. This may include making sacrifices to the river of a goat or a sheep or in accordance with the promises made when the healer procured the necessary medicines. These sacrifices are not part of annual or cyclical rituals, but occur when the believers achieve the outcome sought in the rite. The water spirits at all three falls are part of the same water cosmology, and in the river live innumerable other spirits. On land, there are numerous healers, but only three main healers use water or the river.

The name of the spirit in Bujagali Falls has various spellings. 'Budhagaali' will be used when referring to the spirit and Jaja Bujagali in referring to the medium (although sometimes the healer is also referred to as Jaja Budhagaali). Bujagali is also used for the dam. Jaja Bujagali is the most renowned medium and is said to be the 39th incarnation of the river spirit. However, all healers using the river are challenged by another, Nfuudu, who claims superiority in all matters related to water and water spirits.

Culture and cosmology also represent a wealth of well-being, and this wealth does not always correspond to or sit well with wealth from industrialisation and modernisation. The Bujagali Dam is only some eight kilometres north of the source. The dam was postponed for many years, partly because NGO activists campaigning against it argued that it would destroy the sacred river spirit in the falls and that the indigenous culture and religion would be drowned and disappear in the water. More importantly from a local perspective, the Budhagaali spirit refused to allow the dam to be built, and its medium – Jaja Bujagali – has opposed the dam since the late 1990s. Hence, the spirit blocked the construction of this more than US$ 900 dam for years, along with the corruption that terminated the first phase of the project. Ancient cosmology combated modern technology for quite a while.

President Museveni and the Ugandan government had to incorporate these complaints by international activists and indigenous healers before the dam could be completed. On 19 August 2007, in a state organised ritual, a traditional healer with a spear allegedly moved the Budhagaali spirit from the rapids to another location. This spiritual relocation was successful according to the healer and approved by the president, and paved the way for the completion of the Bujagali Hydroelecrtic Power Station. But it was not Jaja Bujagali – the Budhagaali spirit incarnate – who conducted this ritual. It was Nfuudu, the healer challenging all the others, who relocated the spirit with a bark spear. Without his rituals, the dam project might have been terminated, since it depended on World Bank funding, with all the concomitant social and environmental assessments.

However, Nfuudu was not the proper spirit incarnate and his involvement caused great controversy. According to Jaja Bujagali, the whole 'ritual' was a charade orchestrated by the Christian anti-witchcraft movement in parliament. For years, he had objected to the dam and the very idea that the river spirit could be moved. As he said in one interview: 'If they want to relocate [spirits] to another place, will they carry the whole river or falls to that place? [Do] they really think that a [spirit] is like a goat that can be transferred from place to place? ... The spirits would never allow the dam to be built.'[44] If the dam were built, people would die, he argued. Moreover, he stressed that a river spirit could not be moved except of its own free will. And truly, according to general conceptions after the dam was completed, the Bujagali river spirit is still there and has always been there, and people have died.

'The beginning of Uganda'

Although the Busoga were periodically attacked by Buganda, they were never under Buganda rule. After the British established their protectorate over Busoga in 1894, the Busoga were anxious that their land system should persist, since chiefly political and economic power lay in the land. The traditional authorities were, however, acknowledged as early as 1900. In that year, Commissioner Harry Johnston wrote to the chief administrative officer in Busoga, William Grant: 'We are not here to govern them, but only to guide them and see that they do not commit acts of barbarism and injustice. The policy of the Foreign Office is to govern (i.e., guide) the people through their chiefs and not directly.'[45] The first British post was established in 1893 at Chief Luba's fort. The opening of the 61-mile Busoga Railway in 1912 between Jinja and the Lake Kyoga steamer was a significant step in consolidating Jinja's economic position.[46]

A century after Speke published his *Journal*, 'Jinja represents in some sense an epitome of the tremendous economic and social advances that have taken place in Uganda during the remarkable hundred years since Speke was at Mutesa's court and at the Ripon Falls,' Hoyle wrote in 1963.[47] This was due not only to the railway but also to the Owen Falls Dam. '[W]hat fun to make the immemorial Nile begin its journey by diving through a turbine,'[48] Churchill had written in 1908, and so it was. This was not, however, a straightforward process of development, but included years of hydro-policies and geopolitics in which Egypt's need for water was crucial. The Owen Falls Dam was undoubtedly at the core of the strategy to make Jinja the catalyst for Western investments, thereby serving economic and also political interests that went beyond Uganda. Indeed, from the outset, the Owen Falls Dam was enmeshed in larger world politics. Britain faced a difficult dilemma as regards dam development in Uganda, and the Owen Falls Dam was an act of intricate diplomacy. If Britain listened too much to Egypt, Uganda would be negatively affected, and within the empire some voices argued more in favour of Uganda than others. Yet if Britain supported one colony, it could compromise their position in other colonies as well as raise opposition in Egypt. 'Britain could not alienate Egypt more than had already been the case, and Egypt and Sudan were of much greater strategic importance than Uganda,' Terje Tvedt has written. 'The electrification and industrialisation of Uganda were regarded as very important by the Colonial Office, but at the Foreign Office these were approached as a minor aspect of the much broader issue of Nile development and Nile diplomacy.'[49]

In 1929, Egypt and Britain (on behalf of the East African colonies) negotiated the Nile Water Agreement, which stated that 'no irrigation or power works or measures are to be constructed or taken on the River Nile and its branches, or on the lakes from which it flows ... in such a manner as to entail any prejudice to the interests of Egypt, either reduce the quantity of water arriving in Egypt, or modify the date of its arrival, or lower its level.' The Owen Falls Scheme was part of Egyptian politics, and part of Egypt's 'Century Storage Scheme' on the Nile. The idea was that Lake Victoria, with a surface area of 69,000 square kilometres, could be used as a one-year storage reservoir. If the lake level was raised by only 1 metre, the stored water would increase by almost 70 billion cubic metres, the equivalent of nearly 80 per cent of the Nile's annual discharge as measured at Aswan. The problem, however, was that it would be difficult for this water to flow through the swamps in Sudan, the very swamps that had blocked all searches for the Nile's source for 2,500 years. Moreover, of the water flowing out of the equatorial lakes, half evaporates in the Sudd before reaching Khartoum, where the White Nile meets the Blue.[50]

In the first half of the twentieth century, Bujagali Falls were discussed as a potential site for a dam, but were dismissed as being unsuited to effective control of Lake Victoria. The Owen Falls were preferred. Unexpectedly for the government of Uganda, Egypt wanted the dam to be 1 metre higher than planned and necessary for electricity production, in order to secure and control the lake, and release the water when needed. The Egyptian government paid a total of £980,000 as one-time compensation for the extra expenses and the loss of power production in Uganda.

Although the dam submerged the Ripon Falls, the bar (the rocks of Jinja) that were the geological cause of the falls constrained the flow. These rocks were blasted in 1959 to increase the water flow to the turbines. As with most dams, the initial construction budget was too low. In 1953, the cost overruns amounted to almost 100 per cent. The original estimate was £7.1 million, but by 1953 the costs were £13 million.[51] The final price tag was £16 million. Despite the price, the importance of the dam to Uganda's development cannot be exaggerated: it was as if the country had started anew, and been provided with electricity for decades to come. According to one African journalist, it was 'the beginning of Uganda.'[52]

However, even though Uganda got electricity from the Owen Falls Dam, the discharge of water enabling electricity production was regulated by an Egyptian engineer living in Jinja. Control of water flows was in accordance with the 'agreed curve' and water was released when Egypt needed it.[53] The 'agreed curve' was an agreement that the flow of water from Lake Victoria would be similar to the natural flow at Ripon Falls had the Owen Falls Dam not been constructed.[54] The agreement between Egypt and Britain acting on behalf of Uganda was signed in 1949. Egypt wanted to use Lake Victoria as a reservoir from which it could release 'timely' water during the period when the Nile usually runs low.

The Owen Falls Dam was planned in an era before Aswan High Dam was even thought of, so that controlling Lake Victoria was a vital concern at the time. With the construction of the Aswan High Dam, Egypt lost interest in Lake Victoria as a storage for 'timely' water, since the water supply was now controlled within Egypt's own borders. Importantly, the Owen Falls Dam had strengthened water awareness in Egypt. The long diplomatic struggle over it helped convince Egyptian nationalists that Egypt needed to control the Nile within its borders. It was also clear that a dam on Lake Victoria would not secure Egypt's growing water needs. From this perspective, the Owen Falls Dam was an historical prelude to the Aswan High Dam.[55] Moreover, for Egypt it was a lesson in water security and the perils of leaving control of her life-giving water in foreign hands. These concerns proved well founded. When President Nasser nationalised the Suez Canal in July 1956, Britain planned to use the Owen Falls Dam as a weapon against Egypt. The British considered 'turning off the tap' and thereby harming Egypt downstream. However, this Nile card was geopolitically infeasible, partly because of the complexity of the Nile system. The waters take four months to flow from Owen Falls to Egypt, and although the consequences would be serious for Egypt, there could also be damage to British prestige and devastating consequences for other East African territories.[56] In short, the source of the White Nile at the outlet of the world's second largest lake has been part of the larger scheme of world politics since Speke's time, but has gained in importance with dam construction and Uganda's current role in storing the White Nile's headwaters.

In addition, dams also prompted complaints, then as now, about how they would benefit local people, in this case, the Busoga. Although it was generally acknowledged that the dam would contribute to the betterment of Uganda, there were also vocal critics among the Busoga. In a letter to the prime minister from the office of the hereditary Abataka of Busoga in 1950, the criticism was explicit: 'The construction of the dam and the erection of the electric plant are not being done to benefit the African Busoga nor Uganda. But for the benefit of non-natives in Uganda. For we have no buildings in towns suiting the use of electricity. And more important, we have not been informed of the uses and the purposes for which the Dam and the electric power-plant are being carried out.'[57] The letter went unanswered, and the policy of industrialisation proceeded without the consent of the chiefs and their subjects. In retrospect, the construction of the Owen Falls Dam is no longer seen as controversial, in particular in comparison with the Bujagali Dam, even though the consequences were much the same, including flooding of river spirits.

Powers of the gods

With the construction of the Owen Falls Dam, the waterfalls disappeared. Speke had named the upper falls the Ripon Falls after the president of the Royal

Geographical Society in 1859–60, Earl de Gray (1827–1909), who became the 2nd Earl (1859) and 1st Marquess (1871) of Ripon. However, there were several falls along the river at the outlet of Lake Victoria. A little downstream from Ripon Falls were the Owen Falls. The name Owen Falls seems to have first appeared on Lieut-Col Macdonald's 'Map of Uganda' of 1899–1900, and it is reasonable to assume that MacDonald named these falls after Major Roderick Owen (1856–96). Owen was a close associate of Macdonald, and it seems that the latter literally put his friend on the map. Owen himself came to Uganda in 1893 and there is no evidence that he visited the falls. He only crossed the Nile near Jinja twice, and in a hurry – on his way to and from Uganda. By a curious coincidence, there is also a connection between the two persons the falls are named after. In 1884, Owen was aide-de-camp to the Marquess of Ripon when the latter was the Viceroy of India.[58] Thus, nature in Uganda mimicked society in India, for Ripon's falls are situated higher than Owen's. In the long view, it is fitting that the hydropower plant now bears the name of the water spirit in Lake Victoria in place of that of a passing major.

In the Ripon Falls a particular female water deity lived. Despite the falls being dammed, she is still alive today, just like the Budhagaali spirit. Unlike the other major river spirits, she has not chosen a specific healer as her human embodiment. In Owen Falls too, there were spirits. From a religious standpoint, the constructions of the Owen Falls Dam and of the Bujagali Dam were two very different processes. The building of the Owen Falls Dam was more gradual and cooperative, at least that is the perception today. And the spirits wanted to help the people, and consequently were not wrathful about the damming of the falls. This was not the case with the Bujagali Dam. Although this spirit also wanted to help common people, the project was seen as bulldozing the local populace and the spirits. This provoked divine wrath, whereas if the spirits had accepted the dam as benefiting the people, they would have sanctified the project. In both falls, however, despite their being dammed, all the spirits are still present and active.

When the Owen Falls Dam was constructed, the current Jaja Bujagali was the ritual specialist in charge of appeasing the relevant spirit when the dam flooded the falls, he said. He is at the centre of all waters and everything that concerns water. There is, however, another story. According to Nfuudu, it was his father who conducted the necessary ritual facilitating the Owen Falls Dam. This ritual took place in 1948 and involved the sacrifice of a goat. Thus, there are different accounts, and obviously only one of them can be correct. The involvement of Nfuudu's father in this ritual was challenged and denied by the Busoga chiefdom. Nfuudu and his father do not belong to the Busoga, but to the Luo of the Tororo district in eastern Uganda bordering Kenya. And apparently Nfuudu came alone to Busoga, and hence his father could not have been there at the time. In any event, this is only one of several instances where the stories differ greatly, and these controversies are a recurring theme: who embodies the powerful gods? The healers themselves strongly disagree about who the true incarnates are and these issues are largely decided by the believers and followers. After all, it is they who constitute the cosmology in practice and benefit from the healers as the spirits' embodiments. What is undisputable is that water spirits are believed to be very powerful, and those residing in waterfalls are the most powerful of all.

Today, the waterfalls are also powerful, but generate power in a different way. The Ripon Falls have disappeared under the waters of the Owen Falls Dam and today the hydro powerplant is known by the old name of the spirit in the lake, Nalubaale. The second discharge point from Lake Victoria was completed in 2002 and initially called Owen Falls Extension, but later renamed Kiira. And finally, there is the Bujagali Hydropwer Plant at Bujagali waterfalls. The names of the powerful spirits continue, and perhaps it is more than coincidence that the power of the waterfalls has been incorporated into the secular sphere: the very same waterfalls manifest the powers of the spirits and help power the turbines generating electricity for Uganda. But the secular power of the waterfalls is obviously very different from their spiritual powers.

Bujagali Hydropower Plant

Bujagali Hydropower Plant is some eight kilometres from Nalubaale and Kiira Hydropower Plants (Fig. 5). The project includes 100 kilometres of transmission lines. There were two phases of the project. The Virginia-based AES Corporation was the world's largest private power company. As part of the first phase, AES signed a 30-year agreement with the state-owned Uganda electricity utility, which was obliged to pay AES for Bujagali power even though it was too expensive for most Ugandans. To untangle the knot, the World Bank ordered the Ugandan government to increase electricity tariffs by at least 70 per cent.[59] The first phase of the project was terminated and AES pulled out in September 2003, partly due to a corruption scandal.

Let us start with some facts and figures. Compared to other dams, it is difficult at the outset to see why this dam has been one of the most controversial in recent history, since the area flooded and the number of people resettled and compensated was relatively small and much lower than in most dam projects. The projected price tag was $798 million for the hydropower station and an additional $70 million for the interconnection projects (national grid, etc.). The World Bank provided up to $360 million for the hydropower station and the African Development Bank was involved in loan financing $137 million.[60] A later budget gave the figure of $180 million: $110 million for the construction of the dam and $70 million for the transmission line). All told, there were 10 lenders, including the European Investment Bank. In 2012, the final price tag

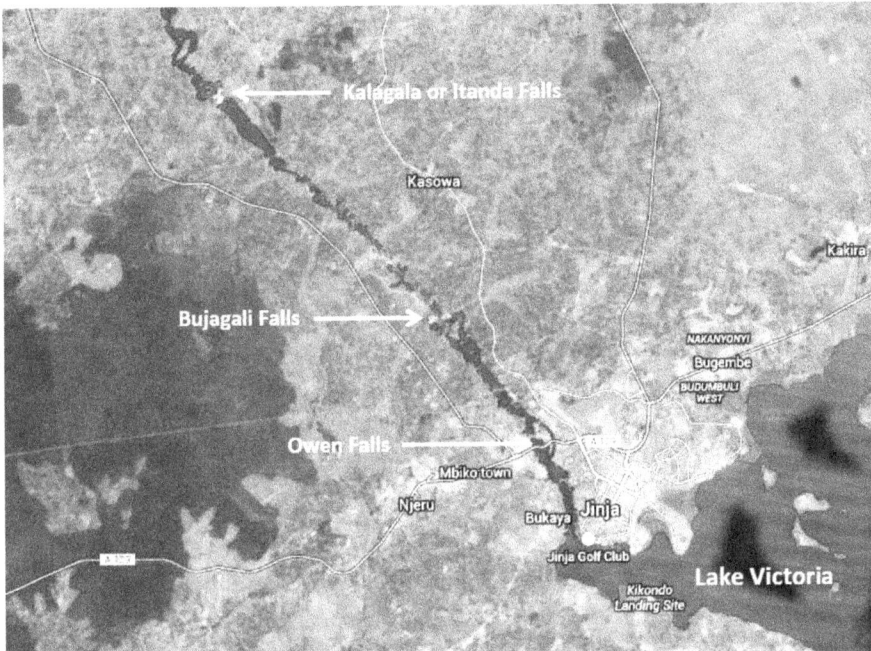

Figure 5. Map of Owen Falls, Bujagali Falls and Itanda Falls. Modified from Google Maps.

was estimated to be $902 million.[61] The bid for the project in 2000 was $447 million.[62] Thus by 2012, the price was more than double than the initial projected cost.

Bujagali Hydropower Project involves a 250 MW hydropower facility comprising a 28 metre high earth-filled dam and an associated power station with five 50 MW turbines. The reservoir extends upstream to the tail areas of the Nalubaale and Kiira dams. The reservoir when full will have a water volume of 54 million cubic metres.[63]

In the Social and Environmental Report produced by the construction company, Bujagali Energy Limited, it is noted that 'the reservoir will be 388 ha in surface area [3.9 square kilometres], comprised of the existing 308 ha of the surface of the Victoria Nile, and 80 ha of newly inundated land. The amount of newly inundated land is small, as the reservoir will be contained within the steeply incised banks of the river.' The same report also refers to 88 hectares of inundated land. The total land-take was 125 hectares, including 45 hectares for the permanent facilities. During construction, an additional 113 hectares was temporarily needed, which was to be restored as parkland, forest or farmland. Thus, the total affected area was 238 hectares. 'At 388 ha, the surface of the reservoir will only be 88 ha greater than the existing 300 ha surface area of the river as it exists without the dam.'[64] The construction was planned to take place over

44 months. The construction and daily operation of the dam were seen as creating job opportunities and economic benefits. Direct employment of Ugandans during construction would range between 600 to 1,100 persons and up to 50 during operation. Induced employment was estimated to be between 9,000 to 15,000 during construction and 250 during the operation, in addition to other benefits in the form of indirect employment and trade.[65]

From Nalubaale power station at the source of the Nile to Dumbbell Island, where the Bujagali Dam is located, the Victoria Nile varies in width from 200 to 600 metres and drops 20 metres in a series of rapids. Thus, the outflow from Lake Victoria has since 1954 been controlled by Nalubaale power station and from 2000 also by Kiira Dam, and has been in accordance with the Agreed Curve.[66] The Bujagali reservoir has limited water-storage capacity and essentially all water released by Nalubaale and Kiira power stations upstream will be used and released at Bujagali. Given the limited reservoir capacity, the water fluctuation in the Bujagali reservoir may vary by up to 2 vertical metres depending on how flow and power generation are controlled.[67]

One concern was that the Bujagali Dam would create increased pressure to release more water from Lake Victoria and subsequently lower the lake level. The construction team counter-argued and concluded that 'the Bujagali dam project will allow for the reuse of the same water that the existing hydro facility at Jinja uses. By reusing this same water, it is possible that the water volume release could be decreased as there would be less demand on the existing hydro electric facilities.'[68] Thus, according to the assessment report made for the team, the Bujagali Dam 'is not expected to significantly alter or affect the hydrology of Lake Victoria or the Victoria Nile … Because the reservoir for [Bujagali] is small it can only hold back a few hours of flow, and therefore it will essentially pass-through whatever flows are released by Nalubaale and Kiira … The only significant concern related to hydrology is concern for the public safety from fluctuating water levels immediately downstream of the dam … Fluctuations further downstream are not expected to be problematic, and not expected to be significant in Lake Kyoga or beyond.'[69] All of these conclusions were challenged by NGOs.

When the first stage of the project terminated, about 8,700 people (1,288 households) had been moved or compensated. Most of those affected were compensated, and only 634 people (85 households) had to move. Of these, 34 households resettled in Naminya, a village approximately five kilometres from the site, and the remaining 51 households resettled without assistance by using the cash compensation. With regard to the transmission lines, fewer people were affected overall, but more were physically displaced. In 2001, it was anticipated that 5,796 people would be affected and 326 households or 1,522 individuals would be displaced. Of these, about 900 people (184 households) would need

Figure 6. The reservoir of the Bujagali Dam.

assistance from the company during resettlement. In 2005, only 27 households had been relocated, and most of them through cash compensation. In the course of the construction of the Bujagali Dam, the transmission line route was optimised, and the number of displaced households declined from the estimated 326 in 2001 to 120 in 2006. The methodology for calculating those likely to be affected or displaced was also a matter of debate. The project contractor reduced the estimated number affected by the dam from 8,700 to 5,158 based on what was termed 'dependents.' This latter group included 'not household members in sociological or economic terms,' for instance children over 18.[70] In the management's response to the inspection panel, a slightly higher number was used. One hundred and one households or 714 persons were physically displaced and the 1,187 non-physically displaced households were compensated for the loss of crops, trees, land and other assets.[71]

Thus, the impact of the dam was slight by any standard of dam construction. Since the reservoir is located in a gorge (Fig. 6), only 80–90 hectares of new land was flooded, and 700 people had to move and 8,000 others were compensated, in addition to those resettled and compensated in relation to the transmission line. If the size and impact of the Bujagali Dam is less than for most other dams, why has it been one of the most controversial dams in history? The answer is not to be found in Uganda, but in Washington and Berkeley and at the World Commission on Dams, which launched its report *Dams and Development. A New Framework for Decision-Making* in 2000. The Bujagali Dam became one of the first test cases to see if the World Bank would follow the new guidelines, which effectively would have put an end to all large dam construction worldwide.

Anti-dam activists and the World Commission on Dams

The importance of dams has been emphasised by many state leaders. The Aswan High Dam was described by Nasser thus in 1958: 'For thousands of years the Great Pyramids of Egypt were foremost among the engineering marvels of the world. They ensured life after death to the Pharaohs. Tomorrow, the gigantic High Dam, more significant and seventeen times greater than the Pyramids, will provide a higher standard of living for all Egyptians.'[72] For his part, Jawaharlal Nehru proclaimed that 'dams are the temples of modern India.' Other heads of state have not gone so far, but political leaders from Roosevelt to Mao Tse Tung to Mandela and Lula da Silva have favoured dam construction as an important development strategy. In rich but arid countries such as the US and Australia, storage capacity is about 5,000 cubic metres per capita, while the equivalent figure in poor countries like Ethiopia and Kenya is only 50 cubic metres (in 2010).[73]

The International Rivers Network (IRN), on the other hand, argues that 'World Bank-backed dams include some of the world's worst development disasters, and their legacy lives on.'[74] This stance is general among anti-dam activists. By the end of the 1990s, it was evident that many dams had had unintended, negative consequences for development and the environment, and many were not even successful in achieving the planned outcomes. Regarding the environmental challenges, the then World Bank President Barber Conable admitted in 1987 that 'the World Bank has been part of the problem in the past.' By 1993, the World Bank had made 527 dam-related loans totalling $58 billion (1993 dollars) and between 1993 and 2002 8 per cent of the bank's lending was for hydropower and irrigation.[75] In 2003, the World Bank's senior water advisor John Briscoe acknowledged that 'big dams account for 10 percent of our portfolio but 95 percent of our headaches.'[76] Big dams, big challenges, but also large opportunities and benefits? Anti-dam campaigners say no.

The commonest water conflicts have not been the anticipated water wars, but conflicts between water developers and dam builders and their opponents, mainly anti-dam activists. From the latters' perspective, the issue is one of watershed democracy and ecology as a human right.[77] Anti-dam activists share certain fundamental premises, which have remained constant: 'that large dams constituted acts of violence against people and nature; that dam builders and financers were not accountable for their actions; that local communities lacked a voice in the decisions affecting them; and that corruption, inefficiency, and repression were inherent in the process of building and operating large dams.'[78] Or in the words of Majot, a big dam 'just doesn't make sense for the 21st century, and no amount of bravado or backlash is going to change that ... In the end, the dam-building industry is a dinosaur: defiant and threatened, saving face, resting on outdated assumptions, and unable to grasp the legitimate economic, social, and environmental factors that are preventing large dams from actually getting built.'[79] Fred Pearce was even more explicit in 1992:

> Modern Engineers portray themselves as the great uncompromising rationalists, harnessing the world's water for the good of humanity. But today, they appear more like antirationalists. Their view of a river as a piece of plumbing does not fit reality. It ignores the great natural wealth of rivers' wetlands, the free irrigation service that many rivers provide on their flood plains. The true rationalists are those who attempt to see the whole picture to view a river as part of a wider world, rather than as a piece of hydraulic engineering.[80]

From 1998 onward, John Briscoe was the World Bank's senior water advisor, responsible for both helping to design the World Commission on Dams (WCD) and facilitate the Bujagali Dam, although he was also criticised for undermining the Commission's report.[81] The initial aim of the WCD was to develop a set of recommendations that would ensure that bad dams were not financed and built and that good dams were built without inordinate delay or cost.

From the outset, the Commission acknowledged that 'the more than 45000 large dams around the world have played an important role in helping communities and economies harness water resources for food production, energy generation, flood control and domestic use. Current estimates suggest that some 30–40 % of irrigated land worldwide now relies on dams and that dams generate 19 % of world electricity.'[82] This basically sums up the arguments of the pro-dam supporters. The two most important reasons for constructing dams are provision of water for irrigation and energy. Irrigated agriculture accounts for only about 20 per cent of the world's arable land, but produces about 40 per cent of global food. Hence, irrigation has been fundamental to global food security and keeping food prices relatively low and stable until the recent financial crises and the peak in food prices after 2008–09.[83] In an era of climate change, hydropower is a green alternative to fossil fuels, coal and nuclear power. Also, given that climate change will be manifested in the hydrological cycle in the form of more drought or floods, storing water in dams is a means of adaptation, as the Commission's report also pointed out. In addition, reservoirs may create increased fishing and transport opportunities, in addition to providing water to ever increasing urban areas.

Worldwide, it is estimated that the total investment in large dams amounts to more than $2 trillion. These investments, however, have social and environmental implications. Some 40–80 million people are estimated to have been displaced by dams (the recent literature most often refers to the higher of these figures), and 60 per cent of the world's rivers have been affected by these dams.[84]

The UN World Water Development Report 2014 reports that dams have displaced 400,000 in Africa.[85] Whereas anti-dam activists use the highest numbers and refer to the global scale by excluding local contexts, the African estimate of 400,000 seems low given that there are currently about 1,000 large dams in Africa.

The Commission focused to a large extent on the negative consequences of dams. In the general debate, it emphasised the:

> … pervasive and systematic failure to assess the range of potential negative impacts and implement adequate mitigation, resettlement and development programmes for the displaced, and the failure to account for the consequences of large dams for downstream livelihoods have led to the impoverishment and suffering of millions … Perhaps of most significance is the fact that social groups bearing the social and environmental costs and risks of large dams, especially the poor, vulnerable and future generations, are often not the same groups that receive the water and electricity services, nor the social and economic benefits from these.[86]

The WCD had two objectives: first, 'to review the development effectiveness of large dams and assess alternatives for water resources and energy development; and to develop internationally acceptable criteria, guidelines and standards, where appropriate, for the planning, design, appraisal, construction, operation, monitoring and decommissioning of dams.'[87] The Commission, however, interpreted its mandate more widely than identifying good and bad dams, because, as it stated, 'the problem of dams [is] a symptom of the larger failure of the unjust and destructive dominant development model.'[88] Thus, it aimed to create a new development paradigm, and dams were the first to be discarded. And although the Commission stated that 'if a large dam is the best way to achieve this goal [significant advance in human development] it deserves our support. Where other options offer better solutions we should favour them over large dams,'[89] this was not the case in practice. In the report it is written explicitly that '… large dams have not helped attain, but rather hindered, "human development",' and the very last sentence summarises the content (or ideology) of the report: the aim is 'to say NO to the perverted development vision, process and projects.'[90] The Commission aimed to achieve this objective by setting out 26 guidelines that effectively would stop large dam construction in the future.

One of the premises, written by Patrick McCully, was that 'a commission of eminent persons independent of the World Bank … must be able to command respect and confidence from all parties involved in the large dams debate.'[91] It was also seen as important to include 'dam critics as central to the legitimacy of the review. This allowed dam critics to wield an unusual amount of power, for without their involvement, the process would lose much of its credibility.' But despite claims of being inclusive, it was acknowledged that the process was highly selective. For instance: 'the exclusion of governments from substantive power in the process was also vital. Had the governments of leading dam building nations like Brazil, China, India, Japan or Turkey formed an organized bloc within the Reference group, it is almost certain that their coalition would have destroyed the Commission's potential to issue a progressive report.'[92]

Of the 26 guidelines, one was particularly important, challenged by states and advocated by the Commission and anti-dam activists. Guideline 3 states that 'free, prior and informed consent of indigenous and tribal people is conceived as more than a one-time contractual event – it involves a continuous, iterative process of communication and negotiation spanning the entire planning and project cycles.' In practice this amounted to a minority veto right.[93]

Dam critics were satisfied with the report. As the International Committee on Dams, Rivers, and People observed, the report 'vindicates much of what dam critics have long argued. If the builders and funders of dams follow the recommendations of the WCD, the era of destructive dams should come to an end.'[94]

The Terminator on the Nile – The World Bank is back

Although the mandate of the Commission was to 'develop internationally acceptable criteria,' the opposite was the case. Moreover, 'the WCD recommendations – grounded in norms of human rights, watershed-scale democracy, and transnational accountability – essentially moved a set of traditional state responsibilities outside the sphere of the state.'[95] As McCully writes:

> The WCD can be described as a globalized and privatized policy process. The public sector was, to a significant extent, marginalized from the process … While this marginalization of the political sector *may seem uncomfortable to those concerned by the ongoing worldwide privatization of former state functions, marginalizing states from the WCD's negotiations does nothing to reduce the importance of states and international organizations* as the main bodies charged with the responsibility of implementing the report's (non-binding) recommendations.[96]

In other words, accountable states were excluded from the process but expected to adhere to and implement the 26 guidelines, essentially by never building dams. From the perspective of governments and the World Bank, these guidelines were seen as extraordinary flights of fancy, and a central issue was whether these guidelines were supposed to be mandatory. '[S]everal of the guidelines were not … remotely practical. Taken as a whole they had never been considered let alone implemented in any country and their demands were so extreme that it was unimaginable that even the most capable of countries could comply with them,'[97] Briscoe says.

The chair of WCD tried to modify the implications of the report, an approach not shared by other members of the Commission and NGOs. 'Our guidelines offer guidance – not a regulatory framework. They are not laws to be obeyed rigidly. They are guidelines, with a small "g", that illustrate best practice and show all nations how they can move forward.' This perspective was dismissed by others, who claimed that the report was the 'Bible', and compliance with the 26 guidelines became the activist mantra for years to come.

The Commission knew all too well that the report would shake the very foundations of the establishment. The agreed process was that a draft report would be circulated to all stakeholders prior to its finalisation. This was not done, allegedly because the authors were working under pressure to finalise the report. However, they were also well aware that there would be a firestorm over the guidelines among many of the stakeholders who had been excluded from the process; if a draft report were circulated, it would never be accepted. The reason was obvious. On the very day the report was released, the IRN declared that 'the WCD recommendations for transparent planning process and "prior informed consent" would virtually preclude future dam projects.'[98] As Briscoe remarks, 'having manipulated a process in which governments (most of them democratically elected) are sidelined, as are the intergovernmental cooperatives (such as the World Bank), "organised civil society" would then use the WCD to ram the recommendations down the throats of those they had successfully excluded.'[99] The WCD chair said of the report: 'If politics is the art of the possible, this document is a work of art.'[100] But the reactions were not what activists and the Commission had hoped. The reception of the report was lukewarm, and at the launch the president of the World Bank, Paul Wolfensohn, noted: 'The critical tests for us will be whether our borrowing countries and project financiers accept the recommendations of the Commission and want to build on them.'[101]

The Commission, NGOs and activists argued that the bank had promised to follow the recommendations whatever they were. The World Bank, on the other hand, denied making such a 'promise,' basically because it would have been impossible, since the policies are set by the board of directors representing the 180 countries that own the Bank. Various countries were highly critical in their responses. 'The Government of China viewed the WCD as very much biased to the developed countries and anti-dam activists and extreme environmentalists. We therefore retreated from WCD in 1998. We think it may be more appropriate to change the title of the report into "Anti-dams and anti-development".' India rejected the recommendations too: 'The recommendations and Guidelines of the WCD are not acceptable to us,' and Brazil stated that 'this would signify, in practice, paralysis in the financing of new dams.' From Nepal there was consensus that the WCD guidelines 'cannot be implemented ... Its recommendations cannot be taken as Guidelines.' Ethiopia hit the nail on the head for many African countries: 'The conclusion of the WCD is totally unacceptable ... If the decision-making framework for future development of large dams is to be used by the Bank it will be unacceptable to nations that have not tapped their water resources.'[102]

Thus, a fundamental feature of the opposition to the WCD report was its challenge to the autonomy of the state. 'Developing countries see it as yet another instance of the imposition on them by the developed countries of an agenda designed in the latter's interests.' The minister of water affairs in South Africa was even more explicit: '[W]e fought for decades and at a great cost to make sure that we have a government which is elected by and accountable to *all* of our people. If we followed the WCD Guidelines we would be surrendering this hard-earned power to groups of NGOs. That we will not do.'[103] The World Bank also rejected the 26 guidelines, but the report also led to the World Bank's 2003 Water Resources Sector Strategy.

> The World Bank believed that adoption of the World Commission of Dams' principle of 'prior informed consent' amounts to a veto right that would undermine the fundamental right of the state to make decisions in the best interest of the community as a whole ... it is not practical and would virtually preclude the construction of any dam ... The World Bank is committed to support its borrowers in developing countries and managing priority hydraulic infrastructure in an environmentally and socially sustainable manner. In doing this, the Bank believes that the World Commission of Dams' Core Values and Strategic Priorities are appropriate principles and consistent with Bank practice and policies. The Bank will not, however, comply with the 26 Guidelines.[104]

The WCD's manoeuvre of excluding governments while at the same time expecting them to accept the Commission's rules 'was an extraordinarily audacious (and in my view, dangerous) process,' Briscoe says, 'since it aimed to substitute the legitimacy of the state (elected in most cases, accountable in all) with the will of self-appointed, anti-dam NGOs that are not accountable to anyone except their fellow advocates.'[105] Moreover, the attempted takeover mobilised developing countries in the World Bank and made them more unified than ever in demanding that their voices and choices of development paths be heard. As Shengman Zhang from China remarked: '... isn't it obvious that the developing world needs dams, and obvious that they need to be done according to reasonable, practical standards?' The World Bank's new water strategy represented a dramatic change: 'now it was not the rich lecturing the poor, but the developing countries who were in the drivers' seat, providing unequivocal support to the strategy. The Strategy was approved unanimously.'[106] In September 2004, John Briscoe, the World Bank's senior water advisor, made a presentation at an international conference in Marrakech. Showing a picture of Arnold Schwarzenegger as *The Terminator*, Briscoe's concluding remark was: 'The takeaway message on the World Bank and infrastructure – We're back!'[107] The Bujagali Dam thus became a battleground for the World Bank seeking to fund dams and anti-dam campaigners seeking to implement the WCD guidelines. And if it is possible to identify one driving force behind the Bujagali Dam in the World Bank system, it is John Briscoe.

Why build dams?

In 2012, the electrification rate in Uganda was only 12 per cent, one of the lowest in sub-Saharan Africa. Even so, this represented a dramatic increase over the preceding decade. On average, in the years 2001–05, 264,000 households were connected to the grid, whereas by 2011 this number was 420,000.[108] Nevertheless, Uganda has had among the lowest per capita electricity consumption rates in the world, at around 75 kWh per year. This is about to change. According to national development plans, the government is projected to install 3,885 MW in generating capacity and per capita energy consumption will rise to 674 kWh by 2015. In the longer run, the aim is to increase installed capacity to almost 42,000 MW by 2040, which will enable per capita energy consumption of 3,668 kWh. This 42,000 MW will include all types of energy, from oil to renewable energy. All told, Uganda has hydropower potential of more than 4,500 MW, 1,650 MW for biomass and 450 MW for geothermal.[109] Although 42,000 MW may sound high, and environmentalists and activists argued with reference the Bujagali Dam that Uganda did not need another 250 MW, these figures should be seen in context. Currently, Norway produces about 29,000 MW of hydroelectricity. Norway is a small country with only five million inhabitants. According to the US Census Bureau, Uganda's 2015 population is 37 million and expected to increase to 75 million by 2040.[110] The need for more energy is indisputable. By 2011, sub-Saharan countries (excluding South Africa) had an installed capacity of only 28,000 MW.[111] In Uganda, dams are mainly for hydropower and a prime development tool.

The role of dams has to be seen in the context of how to develop. According to John Briscoe, the Millennium Development Goals are the greatest setback to development in recent decades. This is not because their intentions are not good – they are, very much so – but because the goals jump to conclusions about outcomes, without sufficiently emphasising how it is possible to achieve them. And this is where, according to Briscoe, water infrastructure plays a fundamental role. All currently rich countries have invested in and developed their water infrastructure, which is a basis for economic growth. Today, rich countries have harnessed more than 70 per cent of their hydroelectric potential, whereas for Africa the equivalent figure is about 3 per cent. The main problem, Briscoe argues, is that the social cart has been put before the economic horse: without economic growth the development goals are impossible, and dams are crucial to this growth.[112]

The UN World Water Development Report 2014 gives a slightly higher figure for developed hydropower potential in Africa, namely 8 per cent. Even this figure is very low, but unlike the WCD, the UN noted in 2014 that 'fortunately, this rapidly growing region of the world has the greatest hydropower potential of any ...'[113] In 2014 John Briscoe was the Stockholm Water Prize Laureate, 'for his unparalleled contributions to global and local water management, inspired by an unwavering commitment to improving the lives of people on the ground.' The Stockholm Water Prize Committee stated that Briscoe 'has combined world-class research with policy implementation and practice to improve the development and management of water resources as well as access to safe drinking water and sanitation.' Thus, it seems the tide has turned and major international organisations are emphasising hydropower and dams. In any event, this is Uganda's favoured strategy.

President Museveni's approach to developing Uganda is industrialisation. For example, he said in 2013:

> No modern country can prosper by agriculture alone. I have told you this many times. These delays for industrial projects can spell a disaster for Uganda if they are not resolutely resisted and defeated. Where will the youths get jobs from? How shall we expand the tax base beyond where it is now? ... On the side of electricity, Uganda has, in the last three years, commissioned the big hydro-power dam at Bujagali that gives us 250 MW [and five other mini-hydro stations] ... This is a total of about 300 MW if you include the 9 MW we are going to get from Buseruka which is about to be commissioned. This is two times the amount of power the British created for Uganda in the 70 years they were here ... I must, again, point out where the real remaining weakness is. This is in the delay of industrial projects ... Without factories, you cannot create employment, you cannot increase the export earnings, you cannot create the market for the locally available raw materials.[114]

Tourism is, however, also a lucrative business. Though tourism is important, today it seems that there are too many hotels and too few tourists in Jinja, and the building of the Bujagali Dam has impacted the business, as dam opponents rightly foresaw. Rafting the falls was a major tourist attraction, but has ended with the damming of the Bujagali Falls. One may still take a boat trip on the reservoir, but this is a lesser attraction. The whitewater industry, should it continue to grow, would generate an estimated $30 million annually, and if hotels, transport, cottages etc., are included, could generate income of up to $60 million a year, anti-dam activists argued.[115] This has not happened. All is not lost, however, for the rafting industry has moved to the Itanda or Kalagala Falls, and large parties raft down the roaring cascades. But if Itanda Falls are also dammed, this business too will wither.

The impact of the dams on tourism is, however, mixed. While the activists argued that they would be a serious blow to local business, other figures indi-

Figure 7. The Bujagali Dam.

cate that tourism has increased. *New Vision* reported in 2013 (20 September): 'Tourism earnings jumped by 22 % in 2011 with the country raking in $832 million, up from $662 million in 2010. This is almost double the $449 million that the country earned from coffee, Uganda's top foreign exchange earner for decades.'[116] Tourism is, of course, much more than whitewater rafting, and has other important implications. Whereas some activists argued that increased tourism should be the preferred development path for Uganda, the government and President Museveni favour industrialisation and energy. Common to both approaches is water as an indispensable resource, whether for recreation and enjoyment or industry and energy.

Summing up the pros and cons (Fig. 7), the World Bank inspection panel noted that 'from a financial point of view, the strategic base for Bujagali remains strong and unchanged; it would optimize productive use of Nile waters ... without increasing the draw from Lake Victoria; it would provide a big extension to the generating capacity ... to cope with growing demand (especially from business); and by the involvement of private companies it would attract (directly and indirectly) both expertise and inward investment.' On the downside, 'an opposing view might be that Bujagali increases dependence on the Nile waters ... compared with other generation options; it pre-empts use of public financial resources; an alternative strategy based on or including dispersed generation through smaller units could more rapidly bring supplies to the un-connected majority of the population, whilst reducing the foreign currency dependency.'[117]

After 30 years, the dam is to be sold to the government of Uganda for $1. As the inspection panel went on to note: 'Once the debt is repaid, the picture changes, in Bujagali's favor. It may then be possible to reduce its tariff. After the first dozen years, it should become a reliable source of cheap power (so long as the Nile flows!) for decades to come.' However, although favourably disposed towards the dam project, the inspection panel had some objections. Private instead of public sector ownership of the project added to the debt burden: 'The high cost of commercial debts in Uganda, coupled with the high-pricing, risk-averting strategy of the investors in response to a perceived high-risk environment, has inevitably saddled the project with large front-end-loaded costs ... As an alternative, public sector financing might have produced lower costs overall, and would certainly have made it easier to manage costs and cost recovery via tariffs over a 40 year project life-time ... it might be argued that a smaller, lower risk infrastructure project would have been a better place to start.'[118] The African Development Bank also concluded that Bujagali 'remain[s] the best solution for Uganda to provide its population with [a] longer-term, lower cost and more sustainable power supply, in response to growing demand.'[119] According to the assessment report for the construction consortium:

> The alternatives to developing Bujagali are to do nothing, or to develop an alternative source or sources of power. The do nothing alternative would mean that the up to 250 MW to be provided by Bujagali would be supplied by extending indefinitely the operation of the expensive high-speed emergency thermals, and by increased load-shedding. This would have a long term significant effect on the economy and the people of Uganda ... The general conclusion from the evaluation of these [nine other] generation alternatives is that large-scale hydroelectric development remains the most economical way forward for the country in the short-medium term. The Victoria Nile is the primary hydrological resource available in Uganda to meet the growing electricity demand in the country.[120]

Accontability, responsibility and representativeness

One of the harshest criticisms of the role of the anti-dam NGOs as campaigners was made by Sebastian Mallaby. In a short article entitled 'NGOs: Fighting Poverty, Hurting the Poor,' Mallaby questioned the responsibility and accountability of the NGOs opposing dams. He set out to investigate the social mass mobilisation in Uganda. In Uganda, he contacted Uganda's National Association of Professional Environmentalists (NAPE), the organisation filing requests to the World Bank inspection panel together with other organisations. NAPE was sponsored by a group called the Swedish Society for Nature Conservation. In Uganda, Mallaby was informed that NAPE was a membership organisation, and then the interesting answer came when he asked 'how many members?' NAPE had 25. He writes:

> Uganda's National Association of Professional Environmentalists had all of 25 members – not exactly a broad platform from which to oppose electricity for millions ... NGOs claim to campaign on behalf of poor people, yet many of their campaigns harm the poor. They claim to protect the environment, but by forcing the World Bank to pull out of sensitive projects, they cause these schemes to go ahead without the environmental safeguards that the bank would have imposed on them. Likewise, NGOs purport to hold the World Bank accountable, yet the bank is answerable to the governments who are its shareholders; it is the NGOs' accountability that is murky.[121]

In visits to villagers who would be relocated, the independent review mechanism found that they were happy to be 'dam people' and with the promised generous financial compensation. Other follow-ups also indicated that the relocated people were fairly satisfied.[122] After inquiring about the life situation for the resettled people in Naminya village, the European Investment Bank reported in 2012 that 'all people present replied that they were doing better now than before and that the quality of life had improved.'[123] On the other hand, the most dissatisfied were those living just next to the project area, meaning that they would not receive compensation.

Mallaby's point is who benefits from anti-dam campaigns, particularly if they are successful in getting the World Bank to withdraw? With the aim of blocking dam building in Qinghai in China, activists enjoyed a fleeting victory when the World Bank pulled out in 2000. But China had no plans to give up dam building and proceeded to ignore the bank's environmental conditions and displace even more people. After the World Bank pulled out, a delegation of Tibetan activists had difficulty in figuring out what was happening, so they asked the president of the World Bank for clarification. Angrily Wolfensohn replied; 'How the fuck do I know what they're doing? You just got us out of there!' Mallaby writes: 'Versions of this story play out all over the world. The bank designs a reasonable project, which inevitably has flaws. NGOs seize on these flaws and add a large sprinkling of inflammatory rhetoric. The World Bank pulls out, but the project goes ahead anyway, minus the bank's social and environmental safeguards.'[124]

Importantly as well, who do the NGOs represent? Civil society will argue that they represent the people or voices that cannot be heard. However, NAPE as an organisation of 25 members cannot be said to be represent anybody but itself. Although a corruption scandal and not the activists finally terminated the first phase of the dam, they certainly did what they could to postpone the whole process. Regarding representativeness, these 25 members and numerous other

NGOs in Uganda and internationally did not represent the millions of potential beneficiaries of the electricity from the dam. Moreover, it is impossible to represent the 'Busoga' as a whole, who number more than 2.5 million and whose opinions are varied. Thus, claims to representativeness are highly problematic, not only because those 'represented' seemingly without 'voices' are diverse and have different and conflicting views, but also because, in varying degree, such claims share the same principle as the WCD, namely that a limited number of people is given a minority veto in practice.

Nonetheless, NGOs and civil society still have an important role and function in democratic and undemocratic societies alike. As Anderson writes: 'The glory of civil society institutions ought to be that they are *not* representative, and, because they are not, are free to argue and shout their visions of social injustice, seek to persuade, and offer alternatives that representative institutions cannot. They, for their part, need to understand that the rest of us are quite free to disagree with their ideas, to confound their plans, and to thwart their actions – or merely to ignore them.'[125] Or to agree. In any event, one has to include the state as an accountable actor regardless of what type of democracy is in place.

All these campaigns have an enormous cost, and even social consequences, when dams are delayed (and other costs if they proceed). According to the contractor, 'it has been estimated in some reports that every month the Bujagali project is delayed costs the economy approximately 10 to 15 million dollars.'[126] The final price tag for the dam was not the initial $500 million, but $800 million ($900 according to recent estimates). Moreover, in 2005, when the country was in the grip of a regional drought, Uganda faced acute power shortages due to delays in the Bujagali project. The government made the decision to procure emergency thermal power generators as a temporary fix, but this was an expensive stratagem that necessitated subsidised tariffs for consumers.[127] It has been estimated that Uganda spent $6 million a month on thermal power generation.[128]

When huge projects are delayed because of the complaints and for other reasons, someone has to pay. According to Mallaby: 'This money comes out of the hides of the world's poor, and the associated delays mean further months without electricity or clean water that a bank project might bring – harming the poor a second time over. The bank's expense and delays do not even benefit the environmental and human rights agendas that NGOs hold so dear. Because of the high cost of doing business with the bank, countries with the option of borrowing on private capital markets increasingly do so.'[129] And here, in particular, China is a key player.

The World Bank and China as dam builders

Geopolitically, there has been a significant shift since 2000 as regards building large dams. Today, about half the world's large dams are in China and the country has a potential of more than 170,000 MW of hydropower. China is also now a major global player, challenging Western domination in dam building. Big dams are big politics, but ironically it was the West that recently provided China with the relevant technology and know-how. China's embarkation on its current programme of big dams more or less coincided with the end of big dams in the West, not because of Western activism, but because most major rivers had been dammed and all large dams had already been built. Consequently, leading hydropower companies such as ABB, Alstom, General Electric and Siemens needed new projects elsewhere. When China decided to build the Three Gorges and Ertan dams in the 1990s, it was still dependent on Western technology and expertise. Rather than being external consultants, Western companies had to cooperate as partners, and by this means the technology and know-how was transferred as well. By about 2003, China became one of the actors in the global market for large hydropower projects. And as with other economic processes in China, it can offer streamlined products at substantially lower costs than Western companies. Whereas the World Bank was only involved in a handful of dams, by 2010 China was involved in at least 220 dams in 50 countries, either as financiers or through its companies. Currently, China is building 19 of the 24 largest hydropower stations in the world, and power-generating equipment is now its second biggest export-earner: only electrical appliances generate greater earnings. With China as a major actor, the World Bank is back in the game again. As already noted, in 2003 the bank reversed its previous caution against large dams.[130] However, for many developing countries, 'China's straightforward approach is an attractive alternative to the endless nit-picking of traditional donors.'[131] In his speech to officially inaugurate the Bujagali Dam on 8 October 2012, President Museveni emphasised again that external funding is not reliable and that Uganda had to be independent of the whims of foreign funders,[132] in this case, Western money.

Moreover, with World Bank-funded dams like Bujagali, the host countries are likely to face delays and much criticism from anti-dam campaigners. Faced with this prospect, many countries favour Chinese involvement, and by defining the dam sites as restricted areas or military zones, silence the opposition and restrict access by demonstrators. In addition, absent World Bank engagement, there are often no external follow-up social and environmental studies. One question that arises, then, is who are the winners and who the losers when the World Bank is excluded from dam building? In most cases the winners are certainly not the poorest and most vulnerable that the WCD aimed to protect.

What, then, is the current status of the WCD guidelines? The Norwegian government's response to the recommendation was an exception at that time: the government suggested that the committee may have 'gone too far in the direction of consensus-based decision-making systems.'[133] At the time the guidelines were launched, the German government went furthest in the other direction by suggesting that there should be a 'world body of eminent people' armed with the WCD guidelines as their Bible, making decisions throughout the world about construction. No government, it was indicated, should be allowed to build a dam unless approved by this board. When questioned about the legality of this approach, the German representative replied that while such a procedure would not be constitutional in Germany, it should nevertheless be imposed on developing countries.[134] This is very much what the activists and NGOs had aimed for. However, state autonomy and democracy mean that this is not the way it should be. The guidelines are for states to follow if they wish to. Uganda and President Museveni do not. Although the WCD was initiated by the World Bank, these guidelines were never World Bank policy, and even if they had been implemented, policies change, as they always have done. In practice, the WCD report is today largely an historical document, an artefact of dam discourse around the turn of the millennium, but not much more. Yet activists still refer to it when opposing dams, as with the Bujagali Dam, but these guidelines have never been part of official policy and most likely never will be. In the meantime, dams continue to be built.

Modernisation or tradition

In *The New Vision* (Uganda) of 6 October 2001, President Museveni is reported to have said that 'Environmentalists are ignorant of environmental issues … there are confused groups of people who know nothing about environment but have managed to block the World Bank grant to AES to construct a power dam here … I am going to write a letter in one of the international magazines … I am tired of ignorant people giving lectures. I will write and rock them!' He added: 'If there is not enough power, peasants will cut down trees to make charcoal, which is the worst environmental threat.'[135] Another newspaper reported on the same day that President Museveni had said that the kind of arrogance that was derailing the AES Nile Power project at Bujagali was the same as that which had created Bin Laden. In *The Monitor* (9 September 2001), Baguma Isoke, minister of state for lands, is quoted as saying: 'People who are trying to sabotage and to stop the development of AES Nile Power are conservatives who do not realise the need for development and change … who want the world to stay as they found it … AES is now the leading investor; over $500m is going to be spent in five years. Our young (people) are going to be employed, we are going to get taxes, and our produce in Jinja is assured of a ready market. Everyone is to benefit from this.'

During the first phase of the Bujagali project, the World Bank functioned as a guarantor, without which commercial and private financiers would not have invested in the project.[136] This provoked much international criticism. 'Unfeasible' was, for instance, the Swedish International Development Agency's characterisation.[137] Nevertheless, with the World Bank guarantee secured, the way for the project to proceed was cleared.

President Museveni officially laid the foundation stone on 24 January 2002, when the dam controversy was at its height.[138] It had taken seven years for the World Bank to reach a decision, which led President Museveni to remark: 'I am ashamed to even come here … I am not happy because a project that should have taken two years has taken seven years to start. All this hullabaloo has been a waste of time and a lack of seriousness … this was a circus.'[139] At that time he did not know that it would take another decade before he eventually would inaugurate the Bujagali Dam in 2012.

The Ugandan minister of energy and mines, Richard Kaijuka, was forced to resign in a corruption scandal. Allegedly, US$240,000 had found its way into his pocket and he had been promised US$260,000 when the contract was successfully signed.[140] Although the allegations were not confirmed, they and other corruption allegations eventually led to the termination of the first phase of the Bujagali project.

Despite this setback, President Museveni was determined to continue with his dam projects at Bujagali Falls and elsewhere. A dam at Kalagala or Itanda Falls had been one alternative discussed before Bujagali was chosen as the preferred site. Kalagala is the name of the falls given by the Buganada on the western banks of the Nile and Itanda the name of the same falls among the Busoga living on the eastern side. Kalagala/Itanda is the next waterfall north of Bujagali Falls. As already indicated, Itanda Falls are also home to numerous powerful water spirits, and in Busoga water cosmology the Itanda spirit is third in importance.

One unquestionable premise of World Bank funding of the Bujagali Dam was that the falls at Kalagala/Itanda should be protected and not used for hydropower. The World Bank was clear that '… the long term protection of the Kalagala Falls and the preclusion of development of hydropower potential at Kalagala is a necessary offset for World Bank Group participation in the project.'[141] One of the main complaints raised by NAPE and others with the World Bank, the African Development Bank and the European Investment Bank, was that they seriously doubted the Ugandan government's commitment not to develop Kalagala Falls in the future, in the absence of a clear and legally binding agreement.

The Indemnity Agreement of 2007 stipulates that 'the obligations of Uganda under this Indemnity Agreement are irrevocable, absolute and unconditional,' and that Uganda shall 'set aside the Kalagala Falls Site exclusively to protect its natural habitat and environmental and spiritual values in conformity with sound social and environmental standards … Uganda also agrees that it will not develop power generation that could adversely affect the ability to maintain the above-stated protection at the Kalagala Site without the prior agreement by the Association.'[142] As part of this agreement, the protection of several forest reserves was also included, among them Mabira Central Forest Reserve. This is the largest remaining indigenous rainforest in Uganda. In parallel with the discussions about constructing the Bujagali Dam in the early 2000s, President Museveni aimed to clear parts of the forest and lease out the area to the Indian Madhvani group for sugarcane production. This caused a stir among the public and environmentalists, as a result of which this forest was included in the protected areas as a trade-off for the social and environmental costs of the Bujagali Dam and the flooding of the Budhagaali spirit. And by coincidence, it happens that the Budhagaali spirit has a twin brother or a twin spirit, also potent and dangerous, and he happens to reside in the Mabira rainforest.

The water spirits existing in most waterfalls are not the priority of the government of Uganda. Bujagali was not the last hydropower project on the Nile, rather than the contrary. The installed capacities of Nalubaale (Owen Falls) Hydropower Station and Kiira respectively are 180 MW and 200 MW. Effective

Figure 8. Itanda or Kalagala Falls.

generation by this complex is lower, however, and ranges between 130–180 MW due to low water levels in Lake Victoria. Bujagali has a 250 MW capacity. In addition to the numerous small hydropower plants being developed and planned, there are plans for several large hydropower projects. These include Karuma (600 MW), Isimba (188 MW), Ayago (600 MW), Murchison Falls (642 MW), Oriang (395 MW) and Kiba (295 MW).[143] And then there is also Itanda or Kalagala Falls, where a 300 MW powerplant is expected to be constructed by 2020 (Fig. 8).[144]

The planned Karuma and Isimba hydropower projects on the Nile are now being implemented. Both are funded by China's Exim Bank and there is additional Ugandan money, and both dams will be built by Chinese companies. The contract for the Ayago Dam is also with a Chinese company. China has promised that the Isimba Dam will be completed within 40 months, in marked contrast with the time it took to complete the Bujagali Dam. The Karuma Dam is expected to be operational in 2018. There are obvious reasons why President Museveni chooses China as a partner rather than Western sponsors such as the World Bank. He learnt his lessons from the Bujagali Dam, with all the delays and activist opposition to dams. In a broader global context, Uganda is part of a growing trend to see China as a favourable partner. Or, in Museveni's words: 'Unfortunately, [during the construction of the Bujagali Dam], we did not have our own money and our partners from outside tended to put frivolous points ahead of the substantive needs of developing infrastructure. Our Chinese friends … have, not only the technical capacity, but financial capacity as well on favorable terms. Chinese lending is also completely free of the usual meddling and high-handedness of some of the friends from outside.'[145] Without explicitly criticising Western donors and their complicated compliance processes, which led to demonstrations, opposition and activism, he clearly felt that matters proceeded more smoothly with the Chinese. In retrospect, the WCD has had unprecedented and unintended consequences: when independent states in Africa have a choice, Chinese partnerships are often seen as more attractive.

Thus, when it comes to dams and politics, it seems that Museveni and the Ugandan government outsmarted all – the World Bank, the WCD, NGOs and the anti-dam campaigners. Uganda has massive plans for hydropower development on the Nile. From Uganda's perspective, it is not a question of which waterfalls will be developed for hydropower: the answer is, all of them. Nevertheless, if Uganda proceeds with the Kalagala/Itanda Dam, in clear violation of the indemnity agreement, there will most likely be more hulabaloo on the Nile. But in this case, it seems the Ugandan government has the upper hand and will proceed: with Chinese involvement Uganda does not need the World Bank. And the activists have been more silent in opposing these Chinese-funded or built dams. That was not the case with the Bujagali Dam. But it is one thing is to build a handful of dams along the Nile, quite another to dam every waterfall in Uganda. If this happens, Uganda may in future regret these development initiatives, not only because there are many development paths and not all are equally preferable at a given time, since technology is advancing, but also because preserving some of the natural beauty and scenery of the Nile is also a form of wealth for generations to come – within and beyond the borders of Uganda.

Inspection panel and water cosmology

One of the main reasons for the opposition to the Bujagali Dam was what anti-dam campaigners saw as a gap between the decision process and the WCD report as the framework for best practices.[146] Frank Muramuzi from the National Association of Professional Environmentalists said: 'Constitutionally the project violates environmental protection rights … We will sue government, World bank or both.'[147] In 2003 F.C. Oweyegha-Afunaduula wrote a short article under the title 'Huge dams as corporate crime and terrorism.'[148] He was the deputy coordinator of Save Bujagali Crusade. In 1999 Oweyegha-Afunaduula was quoted in the Ugandan newspaper *The Voice* (10 May) as saying: 'Bujagali Falls is a cultural, social, ecological, ethical, moral, spiritual and environmental stabiliser for the Busoga, which if implemented would lead to both ethnic cleansing and cultural death of 2.5 million people. Save Bujagali Crusade cannot allow this to happen.' [149] These allegations were repeated in 2004: 'Recent reflections at Save Bujagali Crusade (SBC) and the National Association of

Professional Environmentalists (NAPE) on the decision of the Uganda Government to embrace corporate advice to destroy Bujagali Falls for hydropower have taken seriously the coming assault on the 2.5 million Busoga through the violence, ecoterrorism, ethnic-cleansing and human rights eroding capacity of the Bujagali dam. The view of SBC and NAPE is that Bujagali Falls threatens the cultural, spiritual and ethno-survival of the Busoga'[150]

Although extreme dam opponents made improbable and exaggerated claims, other criticisms were more nuanced. In one report associated with NAPE, the African Rivers Network (Eastern Africa) warns that 'there is a risk of losing credibility if we just continue opposing dams.' Although the network acknowledged that dams have positive sides, it concluded 'it is not necessary to show the good sides of dams because communities are always bombarded with only good things dams will bring, but they do not live to see them. Dam builders are profit oriented and are only interested in the good sides of dams. There was no need to talk about the good things that had been already promised by the developer.'[151] Thus, by definition, anti-dam campaigners most often present only one side of the story, in the same way as dam proponents downplay negative impacts and consequences.

Despite the repeated and media claims by NGOs and activists about the importance of the cultural and religious heritage of Bujagali, their documenting of these fundamental cultural and religious practices is strikingly absent. On the other hand, the World Bank is not renowned for its cultural and religious studies, including in relation to water. However, as regards the Busoga it is in fact the World Bank that has undertaken the most thorough study to date. The bank's inspection panel is an independent accountability body within the bank, with whom those who believe they are adversely affected by a bank project, or believe fundamental principles have been violated, can file complaints. The inspection panel will evaluate whether the bank is following its own policies and procedures by bringing the issues to the highest decision level. The request to critically assess the Bujagali Dam project was submitted by NAPE and other local organisations and individuals in March 2007. A similar request had been made in 2001 as part of the first phase of the project. In 2007, 'the Request contends that the Bank has failed to follow a number of its operational policies and procedures in the design and appraisal of the Project, and that this will result in serious harm to the people living in the Project area and to the environment, in particular the Nile River and Lake Victoria, and to the customers of the generated electricity and to Uganda citizens in general.' The request also identified a number of specific issues of concern.[152]

Although the title of the report does not leap out at the general reader – *Report No. 44977-UG. The Inspection Panel. Investigation Report. Uganda: Private Power Generation (Bujagali) Project (Guarantee No. B0130-UG). August 29,*

2008, – and the language is technocratic, the report contains a wealth of information on the cosmology and the contest that unfolded between Jaja Bujagali and Nfuudu from the early stages of the dam project. The report also gives an insight into the contested process of dam building. The inspection panel sums up the main elements of Busoga cosmology thus:[153]

> a) the spirits are innumerable, powerful and frequently cross over into the world of the living and may do both good and bad, b) they inhabit the same world as the living and are associated with animate and inanimate objects throughout the landscape, c) they can move freely without the need of human permission, d) they have differential power, influence, and interest, e) they are hierarchical, somewhat comparable to the Greek Pantheon, f) they influence the health, well-being and the livelihood of the living, g) more powerful spirits communicate through mediums who do not view themselves as capable of negotiating or predicting spirit behavior – they are mediums of the spirit who possesses them, and h) the mediums are selected by the spirits, not by the cultural (political) leaders.

All the parties – the government, World Bank, construction team, national and international activists, Busoga people, displaced residents, and healers themselves – agreed from the outset of the project that Bujagali Falls were home to a number of spirits ranging from ancestral and family spirits to one of Busoga's most venerated and powerful spirits, the princely Nabamba Budhagaali. In 2001 the panel noted that the leader of the Ntembe clan was Ntembe Waguma and the diviner (*muswezi*) was Nfuudu. A map made in 2001 showed that there were 16 islands, 32 shrines, 10 large trees, 6 rocks, 20 burial grounds, 2 fireplaces and a forest of particular cultural importance in the immediate project area.[154] These cultural sites were to be flooded (Fig. 9).

The Bujagali Falls area included cemeteries, 'islands, sacred groves, rocks, waterfalls, and numerous Busoga spiritual sites. The persistent resistance to disturbance of the site by the Busoga spiritualists and the expressed concerns of the Kyabazinga Institutions is evidence that the Bujagali Falls are a natural habitat of great importance to the Busoga that is being protected by them ... In addition, studies ... suggest a strong ethno-botanical use of the Bujagali Falls project area, in particular the islands, for healing and mental well-being,'[155] the inspection panel notes. Moreover, in one study of the east bank of the river, more than 70 references were made to local medicinal herbs,[156] making the area important for practitioners of traditional medicine.

In the early stages of the process before 2001, the project management concluded that the local community could not see negative impacts on culture and tradition, and if there were any, these could be mitigated using traditional ceremonies. In fact, the management claimed that, based on interviews with 20 focus groups, 83 per cent of the local community did not see the inundation

Figure 9. Drowned islands in the Bujagali reservoir, where the spirits still reside.

as an over-riding cultural and spiritual issue.[157] During the preparations for the project's first phase, local spirit mediums contacted their respective spirits and reported back that if appropriate ceremonies were carried out and financed by the construction team, the spirits would accept the changes to the spiritual and physical landscape induced by the project. In a 2001 report, the dam builders acknowledged that the spiritual area would be largely inundated, but reported that 'it is considered by the parties involved with the spiritual value of the site – namely Nabamba Bujagali, Lubaale Nfuudu and the Leader of the Ntembe Clan that the issue is a local one and the impact acceptable as long as appropriate measures are taken. Toward this end, these parties have given their consistent support to the project, as long as the necessary ceremonies to ensure the appease-ment of the spirits are carried out.'[158]

As part of the studies for the first phase of the project, it was also pointed out that religious matters were very serious for the Busoga. The dangers identified in relation to breaking taboos and disturbing the spirit world included some direct-ly related to construction 'such as machinery injuring workers, breakdowns, and disappearance of livestock, women having miscarriages or producing deformed children, and invasion of the community by foreign diseases and pests.'[159] More-over, divine wrath affected not only local people, and the builders were warned about the possible failure of the project if the spirits were not consulted and ap-proached in an appropriate manner before attempts were made to move them.

Still, in 2008 the inspection panel wrote: 'the full breadth of the Bujagali falls spiritual site at the higher level of Busoga cosmology has not yet been established. At the level of the princely higher spirits, all Busoga clans and their Bujagali associated *baswezi* are stakeholders.'[160] The implications of submerging these spiritual sites, including an island where rituals to find drowned bodies were conducted, were serious for the Busoga: '... the spirits will disappear. Those whose family members regularly use the Nile waters will be put in a situation of fear of the unknown regarding what else to do when one of their people drowns. The associated fear and helplessness, might lead to various forms and degrees of mental breakdown.'[161]

To complicate matters more, many people living in the project area (and subsequently compensated) were not believers in this cosmology, mainly because they were non-Busoga migrants and not the most appropriate people to ask about the significance of Bujagali Falls, whereas most people who believed in the spiritual significance of the site did not live close to the project area. The spirits were for Busoga kingdom, not only for those living close to the river. And although Jaja Bujagali and Nfuudu disagreed about almost everything, on this they agreed: the spirits are important throughout Busoga territory, and the most important of all when it comes to water is Budhagaali.[162]

Compensation and certificates

The Budhagaali spirit has attracted much international attention, for better or worse. In the *New York Times* it was reported in 2001 that traditional spirits were blocking a £500 million dam. Following the media coverage, AES Corp. conducted a study which filled seven volumes, including mitigation strategies for cultural heritage. In the report, a map showed where the numerous trees and rocks considered to be homes of the spirits were located. 'A ceremony must take place at each site to move the spirit and another to introduce it to another resting spot. Each ceremony requires livestock for slaughter as well as local millet and banana brews. Additional ceremonies must take place at each grave site near the dam to properly transfer the remains of ancestors to another location. Two mass ceremonies are also planned to appease the spirit that resides in the water, one gathering for each of the rival clans that claim to be in touch with the Jaja,' it was reported. But as the company's hired specialist in African religions, John Baptist Kaggwa, pointed out, 'Traditional community spirits are not supposed to be transferred ... What we're doing is appeasing the spirits. We're saying to the spirits, "Your home is going to be tampered with. Please allow us to continue".' But negotiating with the spirits, or more practically, negotiating with the healers and those claiming to have a cultural right to compensation, was a challenge: 'Somebody comes in and says this tree is a resident for such-and-such spirit ... Verifying the unknown is very difficult. You have to take the person through many interviews and sometimes you can't disprove them.'[163] In 2001, the Bujagali Project developed a six month programme at the cost of $125,000 for consultation with and compensation of individuals whose graves and shrines would be disturbed, and the appeasement and relocation of spirits.[164] However, the developers felt that in many cases it was impossible to identify with certainty where the graves were, and consequently also impossible to exhume and relocate the remains.[165]

The cultural strategy of the first phase of the project, a line of practice continued in the second, was 'focused on closure, relocating, or appeasing the spirits, compensating when necessary, documenting spiritual appeasement through signed certificates, and setting a finite timeline (originally 6 months in 2001).'[166] Thus as regards cultural and religious matters, the solutions and approaches were practical, technical and mechanical: 'Dwelling sites of spirits important to the local community are being addressed through transfer and resettlement ceremonies. Ceremonies for the Bujagali Rapids have been carried out, although additional activities are being discussed with the Busoga Kingdom. The project will result in flooding of household graves and amasabo (shrines). Where possible these have been relocated as part of the resettlement programme or through compensation payments. Remembrance services to commemorate those buried in the area will be completed. A structure or monument may be erected, either at the site of remembrance or elsewhere, in accordance with the wishes by local communities.'[167]

Although there were innumerable spirits, the main focus was Dumbbell Island and the chief spirit Budhagaali. If the spirits were not appeased, they could become intensely wrathful. The consequences of angry spirits were also emphasised, and this is still an unresolved matter:

> The high spirits (*musambwa*) of one island to be submerged are associated with Kintu and his wife Nambi. They are the founding couple of the Busoga, father of Lubaale and Nabamba Budhagaali spirit. [One] study noted that '*if Kintu and Nambi are annoyed they can* [leave the island and] *come to the land and take domestic animals and even people themselves as sacrifices. No one is allowed to light a fire or burn the bush on the Island. If one does so, Kintu would claim that they are burning his children and can cause harm*'.[168]

Moreover, 'apart from the Nabamba Bujagali spirit, other spirits on the west bank needed appeasement. Resolution of spiritual disturbances is different for clan and family level spirits. Family level spiritual disturbance in the immediate project area appears to have been resolved in the Sponsor's individual mitigation actions.'[169] Thus, the consortium seems to have succeeded well in its efforts to appease the other spirits, in fact giving away money to people not spiritually affected by the dam. When the investors offered monetary compensation, many claimed their share. In 2001, it was reported in a Ugandan newspaper:

Residents of Wakisi, Kikuba, Mutwe and Malindi along River Nile in Mukono District are feasting from money given to them by AES Nile Power, the company which is going to construct a Dam at Bujagali. Most of them were paid handsomely for their pieces of land, which will be affected by the construction of the dam at Bujagali ... most of the villagers, who used to drink local gin known as Nguli, [are] drinking beer. A good number of them have bought vehicles. Most men have bought mobile phones for themselves and their wives. Most of those interviewed by this reporter said they were satisfied by the compensations. They laughed at their Busoga colleagues on the other side of the Nile who are still opposing the construction of the Dam, saying that the place belongs to the spirits. They referred to them as backward.[170]

Appeasement ceremony, 28 September 2001

The management of the dam referred to an 'Agreement for the Mitigation of Cultural Impacts and Appeasement of the *Budhagali Spirit*,' which apparently Nabamba Budhagaali signed on 21 August 2001. The central aspects of these agreements were: 'For the avoidance of doubt, after providing the agreed facilitation for the appeasement ceremonies, the Company [AES] shall deem and *Nabamba Budhagali* hereby asserts that all requisite cultural ceremonies associated with the interest of *Nabamba Budhagali* within the Project Site have been satisfied, the spirits have been appeased ... [the] inundation of the Culture site has been accepted by the spirits ... *Nabamba Budhagali* shall have no more claims for re-consulting or re-appeasing the Budhagali spirit.'[171] It is not clear whether it was Jaja Bujagali or Nfuduu who signed this. In any event, this was only one of many agreements and certificates to be signed in the coming years. Moreover, it also represents one of the many misconceptions during the whole process, since a healer cannot sign an agreement that a spirit will never change its mind.

There was one major ceremony conducted to appease the main spirits, but this seems to have been a charade caught up in fraud and lies. It is worth quoting from the inspection panel report again: 'On September 28, 2001 at the only large ceremony conducted to appease "*the Budhagaali community spirit*" an unspecified number of clan spiritual leaders, the *baswzi abadhagaali* and important dignitaries from all over Busoga were transported to the site at the Sponsor's expense. The followers of Budhagaali were concerned with the rumor that the construction of the dam would take place at their sacred site. They were satisfied, however, when it was revealed that the dam would not be constructed at the site but 3 kilometers downstream at Dumbbell Island.'[172] Neither the dam builder nor the NGOs present corrected this misunderstanding that the sacred sites would not be destroyed, and it seems that the truth of the location of the dam was not explicitly told.

In 2001, the project identified 'Lubaale Nfuudu as a diviner (muswezi) who asserts that the spirit Lubaale is the father of Nabamba Budhagaali spirit. He conducts occasional ceremonies with *buswezi* at the Bujagali Falls to communicate with Lubaale, one of the highest spirits within Busoga cosmology, but different from the Bujagali spirit.' The inspection panel thus concluded: 'This opens the possibility that Bujagali Falls, as a cultural property may be the site of two high spirits of the Busoga, not one.'[173] Moreover, the references to undifferentiated 'Bujagali spirits' makes 'it difficult to determine whether or not there are rival claims or just a rivalry between two spiritual mediums.'[174] While the dam project identified three custodians or diviners, rather than having three separate ceremonies for the appeasement of the Budhagaali spirit, it aimed for one and a co-signed certificate of appeasement, so that the matter could be resolved once and for all, but the diviners refused.[175]

Following the 28 September 2001 ceremony financed by the sponsor to relocate the Bujagali spirits, the builders claimed that all three mediums involved had agreed that the compensation had been adequate and that construction could proceed. They prepared a certificate of appeasement for signing, but at a 2 October meeting Jaja Bujagali withheld his endorsement.[176] According to Jaja Bujagali and Busoga spiritual logic, 'he could not sign the document for the Spirit. He also claimed that the ceremony on September 28, 2001, had been called not to conduct the ritual of appeasement but to consult his *buswezi Budhagaali*.'[177] Jaja Bujagali also handed over a memorandum to the World Bank inspection panel officers signed by 75 spiritual mediums stating that they had never been consulted in the dam process.[178] Moreover, Jaja Bujagali had from the very beginning opposed the dam on the grounds that if the spirits were not appeased, they would wreak havoc on the project through construction problems, and cause death and illness. He concluded: 'The spirits would never allow the dam to be built.'[179]

In early August 2006, the Busoga prime minister expressed concern that the spirits at the falls had not been properly released. At a meeting attended by nine of the 11 Busoga cultural leaders, all of them reconfirmed that the spirits and the shrines needed to be relocated. However, as the inspection panel also noted, although the dam builders to a great extent regarded this *kyabazinga* – or council of the Busoga kingdom – as an important guardian of Busoga cultural tradition, the council did not have the authority to speak on behalf of the spirits: this was the responsibility of the spirits' mediums.[180]

When non-ritual specialists and the secular leadership raised concerns about the appropriateness and effectiveness of the rituals, this was a serious matter and added another dimension to the challenge. The healers are the intermediaries between gods and humans, but the religion and rituals are mainly for the laity. The World Bank also pointed this out. According to the inspection panel in 2008, 'the appeasement ceremony attempted in 2001 organized by Nabamba Bujagali has led to uncertain results. The spiritual medium claims it was incomplete and he is still uncertain whether or not the spirits will be appeased if another ceremony occurs … he has explained that … he cannot predict what the *Nabamba Budhagaali* Spirit will do. Meanwhile, Lubaale Nfuudu has relocated the "Bujagali spirits" to a temporary location, from which they will be moved, once more, to a suitable place away from the Project site to be purchased by the Sponsor.'[181] Thus, there were uncertainties about the rituals and if they had worked, and the whereabouts of the spirit: was it still in the reservoir or in Nfuudu's compound? And was the spirit satisfied?

How much does it cost to finance a ritual paving the way for a nearly billion dollar dam? The amount of money involved in these rituals is difficult to ascertain, more specifically, the amount Nfuudu has been paid to conduct them. On one hand, there was a substantial budget for appeasement ceremonies and religious compensation. As indicated, the Buganda people on the western side of the Nile were compensated well. On the other, Nfuudu told the project's cultural researchers in the first phase of the project that he personally had to borrow money against his land title for a ceremony that required him to transport about 100 *baswezi* (mediums or diviners). Nonetheless, he refused to sign a document assigning the Budhagaali spirit to the Ntembe clan, which according to Busoga cosmology is a contradiction in terms. Following this controversy, he also refused to sign the certificate of completion of appeasement and did not collect the 1 million Uganda shilling cheque for having conducted the ritual in 2001.[182] The contractors agreed that Nfuudu rejected this payment, but as they also pointed out, this was the fifth and final payment for the ceremonies. The previous four payments totalling 12.25 million Uganda shillings for conducting various rituals had been accepted.[183] Where the money went is uncertain. In February 2003, Nfuudu was arrested because he had failed to pay a debt of 1.4 million Ugandan shillings for cement, nails, iron sheeting and other building materials he had received on credit from a dealer. Earlier reports, however, documented that he had received or would be given 20 million Ugandan shillings by AES to support the view that the shrines of Bujagali could be relocated to another place in accordance with the AES environmental impact assessment report.[184] However, even if the contractors had paid the healers for performing a ritual to relocate the spirits, had the spirits accepted the appeasement and the new place?

The inspection panel found that, 'consistent with Busoga belief, ... the spiritual mediums cannot provide assurance as to whether or not the Project could proceed before consulting the Spirits in a manner appropriate to their culture. As Nabamba Bujagali explained ... the Spirit speaks through him. Non-believers may view this response as nonsense, believing that spiritual mediums are speaking for themselves. As such, he can provide no guarantee.'[185] The panel also stressed the importance of recognising 'that mediums of the Nabamba Budhagaali derive their power through the recognition by the traditional clan priests (*muswezi*) as agents of their believers. The mediums of the high Busoga spirits are incapable of commanding their followers ...'[186] Moreover, the spirits decide for themselves if they want to move, and no ritual, regardless of expense and number of animals sacrificed, could change that. After all, gods are gods, spirits are spirits, and humans are humans, and that is the cosmological order.

Moreover, the inspection panel continued (writing in 2008), the 'Bujagali spiritual centrality is not limited to the *Nabamba Budhagaali* Spirit and its me-

dium. Lubaale Nfuudu, a spiritual medium (*muswezi*) for another of the princely high status Busoga spirits, Lubaale, has temporarily relocated some Bujagali
spirits. Lubaale Nfuudu takes care of multiple shrines (*amasabo*) where new
spirits were arriving all the time … Whether or not he relocated the *Nabamba
Budhagaali* Spirit is unclear and probably immaterial, since the Spirits are free
to move wherever they wish.' In addition, 'the *Ntembe* clan, whose leader is
Ntembe Waguma, and diviner (*muswezi*) is Nfuudu, see the Bujagali Falls as
the location of their clan level ancestral spirits which will be disturbed by the
project. Lubaale Nfuudu is the caretaker for his clan spirit and Lubaale, another
Busoga ancestral spirit. He also states he built Nabamba a shrine and questions
the legitimacy of Nabamba Budhagaali as a medium.'[187]

This controversy revolves around one fundamental question. Was Jaja Bujagali or Nfuudu the true Budhagaali spirit incarnate? Which healers had the
spirit Budhagaali chosen to speak through? Both healers fiercely argue in their
own favour, Jaja Bujagali arguing that the spirits would not allow the dam to
be built and Nfuudu claiming that he could relocate the spirit in order for that
to happen.

One thing is certain: He who is the proper incarnation is the most important
and powerful of all these healers. The Nabamba Budhagaali spirit 'can possess a
spiritual leader from any Busoga clans, who becomes, according to interviews,
"like and archbishop" among the clan spiritual leaders … Each clan can have …
a clan level spiritual representative [priest] of the high priest who is ordained at a
sacred rock associated with the high spirit at the Bujagali Falls religious site. The
recognition and initiation of *Nabamba Budhagaali* Spirit's medium is presided
over by a conclave of these representatives, jointly known as *Baswezi Budhagaali*.
Presently, Nabamba Bujagali is the medium for the Nabamba Budhagaali Spirit.
His initiation was recognized by Busoga clan spiritual leaders (*baswezi*) and
other seers.'[188]

So what was the relationship between Jaja Bujagali and Nfuudu? As noted,
Jaja Bujagali was seen as the proper incarnation by fellow healers and diviners
as well as by common people. Why was his authority challenged by the govern-
ment and Busoga kingdom? Another inspection panel report, with an even more
impenetrable title, *Report No. 23988. The Inspection Panel. Investigation Report.
UGANDA: Third Power Project (Credit No. 2268-UG), Fourth Power Project
(Credit No. 3545-UG) and Bujagali Hydropower Project (PRG No. B 003-UG),
May 23, 2002*, states:

> The Nabamba has what is essentially a self-selected position. At some point,
> he says forty years ago, he became aware of his spiritual calling as the embodi-
> ment of the Bujagali spirit, and subsequently had his claim recognized by a local
> following, even though he never completed the traditional course study of a

diviner or 'muswezi.' The fact that he is not a muswezi has been raised by some of his detractors. He is registered with the Government as a 'native doctor.' His residence cards for the district date back to 1988. Lubaale Nfuudu occupies a somewhat more bureaucratically defined position. He, too, is a 'native doctor' with what appears to be a very substantial practice. More importantly, he is the official diviner of the Ntembe clan ... which traditionally 'owns' the Bujagali cultural site.[189]

Obviously, this is a misleading or mistaken description. There is no such thing as a 'traditional course study' for a diviner. Either a spirit in-dwells in a human being or it does not, and no 'study course' can change that. However, a novice may undergo a long initiation before becoming a traditional healer. Equally, a spirit may choose anybody, even someone without prior skills. When this happens, the initiate goes through a long and complicated installation process, which is supervised and acknowledged by a group of other diviners. And Jaja Bujagali had done that, as all agreed. He also had official documents and a certificate with signs and seals acknowledging him to be the Living Bujagali, documentation he had shown the World Bank.[190] Nfuudu's position within this cosmology is more uncertain.

In any event, Nfuudu Lubaale felt that the spirits had been temporarily relocated to his place, and would then be relocated again to shrines near the project site. Nfuudu argued that the spirit Lubaale, one of the high spirits in Busoga cosmology, is the father of the Nabamba Budhagaali spirit. As such he conducted several rituals and consultations with Lubaale at Bujagali Falls. Although perceived as a father, the two spirits are nevertheless distinct. Indeed, there are innumerable spirits at the falls, and Lubaale is not strictly a water deity in Busoga cosmology, although Nalubaale is a water spirit among the Buganda. And as will be evident later, even though these spirits have many family relations or constellations, perceived fatherhood does not imply patriarchal dominance or superiority in determining offspring spirits' agency. The spirits are autonomous and quite independent, despite 'family' relations. Moreover, the strength and importance of spirits is not determined by their ancestry as such. In any event, neither the World Bank nor the dam builder had sufficient knowledge of spiritual matters at Bujagali. The inspection panel concluded in 2008 that the 'project failed adequately to consult with the Busoga spiritual clan leaders associated with one or more high status Spirits about the significant cultural patrimony of the Bujagali Falls.'[191] Moreover, it 'publicly injected the [World] Bank into a religious misunderstanding without competence in the spiritual context of its position, including passing judgment on the legitimacy and credibility of a spiritual medium's performance. *Management unnecessarily and inappropriately took sides in a spiritual controversy of a religion in which millions of Ugandans believe.*'[192]

According to the inspection panel, the dam builder not only misinterpreted the importance of the project site to those living next to the river, while disregarding its importance to the Busoga kingdom, but also omitted to consult some 340 Busoga clan spiritual leaders with spiritual and cultural ties to the affected area.[193] The inspection panel also criticised the dam team for losing objectivity and taking part in an internal conflict, not surprisingly on the side that favoured them, by claiming that:

> ... there has been a fierce rivalry between Nabamba Budhagali on the one hand and Ntembe and Nfuudu on the other during the whole consultation and negotiation process, Nabamba has been quite successful *in attracting media attention and obtaining significant compensation*, whereas the other two seemed to be *more genuinely interested in cultural and spiritual* aspects...[and with reference to Jaja Bujagali they stated that] while the two other stakeholders appear to have *been genuinely satisfied* with measures taken by AESNP [the dam builder], the Nabamba Budhagaali medium seems to have remaining claims over the site. This particular individual has been able in the past to draw a lot of attention, including international attention, which later *did not appear to be justified by his actual spiritual performance, in contrast with the other two. It cannot be excluded that he will seek to obtain more compensation* through media coverage for instance.[194]

An independent review concluded that this failure to reach agreement is significant since 'Nabamba Bujagali and his followers were entitled to particular attention because of the profound importance of the Bujagali Falls as a religious site; the enormity of their potential loss when the Falls are flooded; the impact of this loss on their beliefs and religion...'[195] All these issues could have had a major impact on the decision to choose the Bujagali site as the best location for the dam. 'The Panel notes that if the Busoga religion and cultural tradition had been ... more fully understood and widely recognized ..., the current site may not have been acceptable, or alternative sites would have been given ... much stronger consideration.'[196] In other words, if the World Bank had properly understood the complicated processes at work at Bujagali Falls, it might have refused to support the Bujagali Dam and the project might have been terminated once and for all, or at least until the Chinese entered the scene. However, neither the World Bank nor the anti-dam campaigners knew enough: in fact, the latter did not bother to document the heritage they claimed to protect. The inspection panel report of 2008, when the dam was already under construction, came too late to stop the project. On the other hand, it seems that President Museveni was well aware of the importance of the Budhagaali spirit in Busoga cosmology, and that the religious 'problem' had to be resolved in an efficient manner. The solution was to orchestrate another ceremony, again with Nfuudu as the maestro.

Relocation ceremony, 19 August 2007

Despite the termination of phase one of the Bujagali Dam on account of corruption allegations and all the challenges by NGOs and resistant water spirits, President Museveni was determined to build the dam. In an address to parliament on 10 February 2012, he explained:

> The strategic mistake on electricity took longer to cure. It was only in 2006, soon after the elections, when I proposed the setting up of the Energy Fund, using our own money, that we began addressing that mistake. Previously, we had relied on endless begging from outside for funds to build the dams. The begging itself was further complicated by the internal sabotage by the indisciplined political elements whom the external benefactors were only too eager to listen to. By setting up the Energy Fund, we were now able to build these dams ourselves if it was necessary – although we continued to welcome external financing, if available. The usual games did not take long to manifest themselves. Some old man [Jaja Bujagali] claimed to be the personification of the *'Bujagali spirit'* and 'the spirit was objecting to the building of the dam.' Some of the lenders started procrastinating using frivolous excuses – they must consult civil society, etc. Who is 'civil society'? These are some individuals in the employ of foreign NGOs. How can these be 'civil society'? I understand civil society to mean people who are living by their own means – not employed by Government. This means industrialists, farmers, hotel-owners as well as workers in the private sector, etc. Why should people working for foreign governments that fund the NGOs be regarded as civil society? Fortunately, by this time, a contractor had already been selected. We, therefore, released US$75 million to them from the Energy Fund as bridge financing to start building the dam straight away. If they would not start, we would kick them out and build the dam ourselves using our own money. That is how Bujagali started on schedule this time.[197]

However, even though the president had the necessary financial backing, he still had to resolve the spiritual issues. The Busoga kingdom, the World Bank and other agencies agreed that the outcome of the previous ceremony in 2001 was uncertain. Moreover, given the preceding history, it was clear that World Bank money depended on a satisfactory resolution of cultural heritage matters, and it was also clear that activists would again use the Budhagaali spirit as an argument for blocking dam construction. With the Kalagala offset as a bargaining chip, and with World Bank money secured for the second phase of the project, there was only one invisible spirit hindering the process. The issue was to be settled once and for all by yet another grandiose ritual whereby the obnoxious spirit would be removed, thus enabling the construction to proceed. The ceremonial master of this state-sponsored ritual was the diviner Nfuudu.

The appeasement ceremony on 19 August 2007 was conducted only two days prior to the official groundbreaking for the dam, and as such cleared away the last hindrance to the Bujagali Dam. When President Museveni again laid the foundation stone during a ceremony on 21 August 2007, he said: 'You cannot claim to be protecting the environment when you are denying over 90 percent of the population access to electricity.' Development and industrialisation were the priorities, not culture and indigenous religion.

A Ugandan newspaper characterised the ritual as 'bizarre.' '[T]he disturbing question was how the spirits could be resettled since they were not like the Internally Displaced People in northern Uganda who at least are given seeds, pangas and hoes to start a new life.' Moreover, 'unusual as it may appear, the Busoga traditional healers bulldozed it and performed rituals at Bujagali in Jinja to relocate spirits from the falls … [However] the spirits had earlier already been "removed" from the falls and kept by Jajja Nfuudu … only waiting to be resettled.' The devlopers had bought a new piece of land in accordance with the compensation agreement. '[T]he healers were called to perform at the exercise and one, Nfuudu was given all the necessary requirements which included, goats, sheep, cowries, clay pots, and beads to appease the gods … Nfuudu was mandated to take measurements for the shrines at the new site with respect to the cultures of the land as recommended by the Busoga Kingdom … the Kingdom officials prefer working with Nfuudu because the spirits "acknowledged" him.' Still, not all the hereditary chiefs were satisfied with the ritual, 'arguing that the spirits will be missing the chance of "enjoying" the water … they only agreed to relocate the spirits but not the site [suggesting that] the spirits would have been asked to choose a site because the eight kilometres from the falls [to the new shrines] seem like chasing the spirits from the water where they were "created".' The representative of Bujagali Energy Limited (BEL), on the other hand, Dr Florence Nangendo, was satisfied: 'We are here to resettle the spirits so that the dam can go on. The project respects cultural sites and cultural beliefs.'[198]

Jaja Bujagali was not satisfied at all. Not only had he deliberately not been invited, and hence not been there when the ritual was conducted, but as the Budhaagali incarnate proper, he would never have allowed this ritual to take place. In spiritual terms, the spirit would have informed Jaja Bujagali that it did not accept the ritual and that it was furious at what had happened. Then Jaja Bujagali would have conveyed the spirit's message to the world. Indeed, this was what Jaja Bujagali did. The next day he announced on a local radio station in Jinja that he had never authorised the relocation and no agreement had been reached with him.[199]

The transfer of the spirit was symbolic. Nfuudu placed a spear wrapped in bark cloth in the roaring waters of Bujagali Falls, holding the spear there for some time before pulling it out and later taking it to the new site at Namizi West, some eight kilometres from the falls, where the new shrines were not completed. This was the final relocation of the spirits, the healers agreed. 'The spirits have accepted to relocate. To prove that they were happy, there was rain as we relocated them. This will allow the construction and completion of the Bujagali hydropower project successfully,' said James Christopher Mutyaba, the leader of the healers. The chairman of the Busoga chiefdoms sacrificed three cows, 20 goats and chickens at the three new temporary shrines, which had been hurriedly built of brick. 'The blood sacrifice is to get the spirits embedded under the River Nile waters to relocate. This is a clear testimony that we are behind the project.' For three days the festivities went on: 'they sang, feasted on meat, matooke and drank local brew during the rituals that kicked off on Sunday and ended yesterday. Dressed in bark cloths and beads, joyful elderly men and women numbering about 30, smoked tobacco in long brightly decorated pipes as they danced at their new site worth over sh 11m. The whole relocation exercise cost about sh 21m. Project developers, Bujagali Energy Limited, purchased the site measuring 1.2 acres, following a compensation agreement.' [200]

However, at that time the three new shrines at Namizi were merely temporary structures, and there was also only one substandard toilet, and it was said that the spirits could not move to unfinished shrines. The road to the shrines was in bad condition and impossible to use during periods of rain, and there was no electricity or water facility at the shrines. BEL had released 21 million Uganda Shillings for the appeasement and relocation of the spirits, which included buying land for the shrines in the village. Nevertheless, since the shrines were not completed, the money was spent on relocating the spirits to Nfuudu's place. 'Even though the consumables at the time was not worth what was budgeted for, the fact remains that the food was consumed and a plan needs to be made to address the issue at hand,' the permanent secretary of the Busoga said. 'The spirits have not reached their final destination. The shrines have not been completed because funds were not ample ... According to the Busoga Kingdom, several items that were crucial were omitted in the previous budget ... It was resolved that the Kingdom submits another budget to complete the construction of the shrines ... [and] it was also agreed that the Busoga Kingdom presents accountability of previous expenses for construction of the shrines and the relocation ceremonies worth Ug Shs 21 million that was given to the Kingdom by BEL,'[201] an NGO witness report noted.

Although the spirit was allegedly moved to the Namizi shrines, for most of the time it resided at Nfuudu's compound, if one is to believe the healer (Fig. 10). In my discussion of this matter with Jaja Bujagali, he said he was in Kampala with his son at the time Nfuudu conducted the ritual allegedly transferring the spirits. They watched the coverage on television. Jaja Bujagali has long hair and dreadlocks. Nfuudu also had long hair then. According to Jaja Bujagali, there were two main healers conducting this mock ritual, and Nfuudu stole his identity by claiming that he was the proper Jaja Bujagali. This was possible since

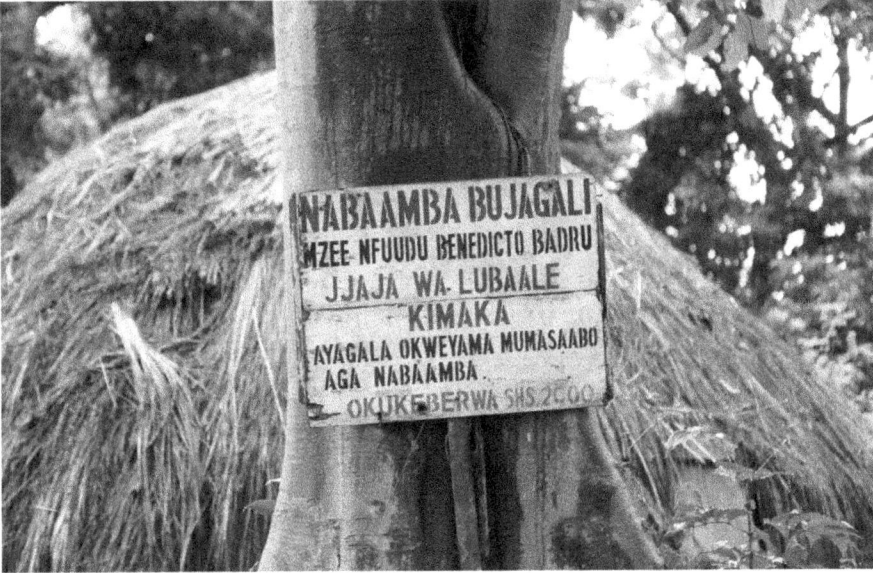

Figure 10. Visitors in Nfuudu's compound reading his sign: 'Nabamba Bujagali. Mzee Nfuudu Benedicto Badru. Jjaja Wa Lubaale Kimaka.'

Jaja Bujagali is not supposed to cut his hair, and the resemblance convinced people that Nfuudu was the proper incarnation of the Budhagaali spirit. Today, Nfuudu has boldly shaved his hair. If a healer cuts his hair improperly, this is a grave act with serious implications for himself or others.

This theft of identify prompted Jaja Bujagali to file a case with the police. Nfuudu was arrested and the case went to court, but there have been no further repercussions for Nfuudu. With this charge, Jaja Bujagali and other traditional healers wanted to expose Nfuudu's bad attitude and to make a statement about his actions and behaviour. The Civil Investigation Directorate received the file on 28 April 2011, and Jaja Bujagali showed me the receipt.

A crucial ingredient in the conflict was, as often is the case, money. Jaja Bujagali is renowned for being a tough negotiator and difficult to bargain with. He had clear demands and uncompromisingly argued that the mandatory ritual ceremonies had to be performed. This included the planting of 700 new sacred trees for herbs and medicines, since Dumbbell Island would be flooded. Moreover, the rituals he demanded were complicated and expensive. When the negotiations stalled, the kingdom chose Nfuudu as its representative healer, since he was more cooperative and easier to negotiate with. In other words, using Nfuudu meant paying less for the rituals. Moreover, the kingdom used Nfuudu as its healer to get money from the World Bank, according to Jaja Bujagali.

Jaja Bujagali's main and non-negotiable demand was that a major ritual should be performed that would – or perhaps could – secure the spirit's blessing for the dam project. Dumbbell Island was the epicentre of all the spirits residing in the Nile and a national shrine for all Uganda. A ritual ceremony was required that would unite the Busoga and even the country, according to Jaja Bujagali. This ritual would have included more than 4,000 delegates or healers from all over Uganda. In a cultural heritage perspective, one can only imagine what a ritual involving several thousand healers and subsequent sacrifices would have been like. However, this ritual never took place, and most likely never will, even though Jaja Bujagali insists that the proper rituals have yet to be performed.

At the outset, such a large-scale ritual may seem megalomanic and excessive – in fact impracticable and too expensive. No wonder the kingdom and government opted for Nfuudu's smaller and cheaper ritual. However, this may not be a fair assessment for one important reason. The money for the compensation rituals was already there. Donors and the contractor had allocated money for these purposes. According to Jaja Bujagali, the money initially intended for the major ritual was misappropriated and misused, and disappeared during the first corruption scandal. Originally, 1.6 billion Ugandan Shillings had been earmarked for ritual compensation. Of this, less than 100 million Shillings were given to ritual compensation, and the rest disappeared. Hardly any of this money was given to Jaja Bujagali to conduct the necessary rituals. On the other hand, according to Jaja Bujagali, Nfuudu was paid 41 million Uganda Shillings for performing the 2007 ritual.

Most of the money allocated for compensation was misappropriated. Everybody claimed to be eligible for compensation, but large sums disappeared through corruption, Jaja Bujagali said. Even national environmentalists received money from the contractor. Some of these environmentalists claimed that they were in alliance with Jaja Bujagali, but without his knowledge subsequently claimed compensation. After that, he never saw them again, and consequently Jaja Bujagali felt deceived by the activists who had used him to stop construction of the dam, but did not care about the spirit.

If the money had been used for its intended purpose and the ritual organised and orchestrated by Jaja Bujagali himself, today, Jaja says, that the Budhaagali spirit would have been pleased and the dam construction and the flooding of the spiritual sites could have been allowed. If the originally planned rituals had been conducted and if the dam had improved the lives and livelihoods of common people, then the construction of the dam would have been beneficial. As Jaja Bujagali rhetorically asked, who is against development?

Final relocation ceremony, 21 June 2011

There was one last ceremony in this ritual drama. The European Investment Bank also contributed loans to the second phase of the Bujagali Dam. As with the World Bank, the European Investment Bank has an independent complaints mechanism. And as with the World Bank, it was NAPE together with other organisations that filed complaints that the bank had violated its own policies and practices. This complaint was submitted 2 December 2009. According to the standard procedure, the European Investment Bank Complaints Mechanism (EIB-CM) evaluated the complaints and published its findings in a separate report.[202] One complaint concerned the 2007 ritual, and rightfully argued: 'Cultural and spiritual resettlement: No proper consultation ever took place with Bujagali and the spiritual community of the Bujagali Falls. Instead, Jaja Bujagali was marginalised from the process, and a fake resettlement ceremony was organised with the complicity of the Government of Uganda. As a result, no proper spiritual resettlement ever took place.'[203]

The spirit Nabamba Budhagaali and its medium Jaja Bujagali 'are unquestionably tied to the 338 Busoga clans,' and the spirit can possess a spiritual leader from any of these clans. Moreover, as the EIB-CM emphasised, 'notwithstanding all the items surrounding the spiritual issues it is clear … that Jaja Budhagali was recognised before as the rightful spiritual leader and that somehow he needs to be taken into consideration in order to progress … in the best possible way.'[204] In response to the complaint, 'in January 2011, a tripartite "Agreement for the Final Relocation and Appeasement of the Bujagali Spirits" was signed between the Busoga Kingdom, the Government of Uganda and BEL. This agreement defines the role and responsibilities of the … parties involved … in terms of the (i) construction of shrines and (ii) organisation of the appeasement ceremonies and ensures the involvement of all key spiritual mediums, including Jaja Budhagaali and Lubaale Nfuudu. The agreement also accounts for completion undertakings to avoid any future claims regarding spirit relocation and appeasement.'[205] In a letter of 11 February 2010, the prime minister of the Busoga kingdom states that Lubaale Nfuudu is the recognised spiritual medium of the kingdom and declares that the Busoga kingdom 'shall accommodate Nabamba Budhagali and make him accept the status quo.'[206]

Not surprisingly, although it was acknowledged that Jaja Bujagali had been bypassed and NAPE characterised the 2007 ritual as fake, the intention of bringing both healers together to finally settle their disputes failed again. Jaja Bujagali did not respond to the kingdom's invitation and refused to participate. Not only would the ritual be futile in his opinion, but he did not want to be treated as the equal of Nfuudu. On the contrary, Jaja Bujagali wanted to conduct his own ceremonies, which would involve very many spiritual healers from across Uganda at a high cost.[207] As a consequence, in the absence of Jaja Bujagali, the ritual proceeded with Nfuudu. If the 2011 ritual was necessary because the 2007 ritual was unsatisfactory or a fake, one may characterise the 2011 ritual in the same way.

At the end of June 2011, Nfuudu again moved the Budhagaali spirit. The

Figure 11. The Namizi shrines when they were finished, but lacking spirits.

spiritual contestation between Jaja Bujagali and Nfuudu continued, and still does. In a newspaper, it was reported: 'The move, carried out last week, ended an impasse that had delayed the ritual for over three years … The conflict resulted in a stalemate until…Nfudu secured the blessings of the kingdom officials to conduct the event. The spirits were moved from the dam site to Namizi East village in Budondo sub-county in Jinja.'[208] Although Nfuudu claimed that the spirit resided in his compound, the ritual started by the river or the reservoir. As before, it was transported back to the new shrines in Namizi village (Fig. 11), the place used in 2007 when the shrines were temporary. As the spirit was being moved for the third time, one must assume that it had been rather dissatisfied with the new shrines and at being moved around.

'We hereby certify that the construction of the shrines and associated features has now been completed to our satisfaction …,' states a certificate of Completion of Namizi Shrines and Relocation of Budhagali Spirits, signed by the Busoga kingdom and the government. 'Today is an important day for Busoga and Uganda. Giving up our shrine [the waterfalls] was not easy – but for development's sake, we had to. The government should give our children jobs because the power project will benefit all of us. Busoga Kingdom should also get its fair share of the proceeds from the project,' the acting *Kyabazinga* (king) of Busoga kingdom, Kawunhe Wakooli said. Benedicto Nfuudu, the chief spirit medium, conducted the ritual, and while standing in front of a fire from which smoke arose, he declared: 'Ever since the spirits were temporarily kept at my home after the ground breaking ceremony for the power dam, I have not had peace. They have been nagging me for another home ... These new shrines are in honour of the Busoga. People will be coming here to worship the spirits of our dead who protect, direct and guide us. I pray and ask that the Busoga sacrifice a cow each year to appease the spirits.'[209]

The contractor was also appeased. 'In as far as the transfer and appeasement of the Budhagali spirits was concerned, we had difficulty in finding an alternative home. BEL was required to find a new home with guidance from Busoga Kingdom. We thank the Busoga for giving up their land for this project because the benefits of the power project are greater than their personal interests,' BEL's head of community interventions, Zakalia Lubega, said. 'All stakeholders, including those who had an interest in the Budhagali spirits, were compensated and cultural ceremonies such as re-burial of the dead and relocation of household shrines and spirits performed ... However, the final resting place for the spirits had not been completed, hence the importance of today's ceremony.' The hereditary ruler of Butembe chiefdom and custodian of the Budhagali spirits, Chief Waguma Yasin Ntembe, said: 'We are grateful to the government for being keen on the importance of the Budhagali spirits. If this was not done the outcome could have been devastating.' He added: 'Whenever you talk of a kingdom you are talking about the spirits, gods, name it. So the kings are the overall administrators of the spirits. Today we are transferring the spirits of our grandfathers. Budhagali was a person, we should remember in the same vein Christians hold memorial services for the departed relatives.'[210]

Prior to the ritual, financial matters were once again a cause of dispute. Although the kingdom had promised to account for how the money had been spent during the construction of the unfinished shrines in 2007, they had not yet been able to do so satisfactorily. BEL argued that it had already paid out 21 million Shillings for the rituals. Where, it wondered, had the money gone and how had it been spent, since the kingdom now argued that it needed additional 39 million Shillings? Eventually the kingdom presented accounts of how the money had been used, and the builder provided an additional US$ 20,000 for the ceremonies to finalise the relocation of the spirits. The role of the contractor in this process was highlighted by the European Investment Bank: 'The behaviour of BEL may be considered as very positive and accommodating as they did provide funds and are in fact the only party to do so regardless of the fact that they are the least responsible for these issues.'[211] Although the contractor had an obvious interest in bringing closure to a dispute that had dragged on for more than a decade, it also had been caught up in a religious fight about which it had insufficient knowledge and which it tried to resolve according to best prescribed practice, namely issuing certificates aimed at settle religious questions. Religion does not work that way, but in its technocratic fashion the contractor tried at least to do its best, and the studies during the first phase before 2001 were meticulous in documenting the spiritual sites affected by the dam.

The financiers and the contractor were satisfied – the spirit finally had a new permanent home. But upon my visiting the shrines in February 2013, there were no signs of any spirit residing there. The alleged destiny of this mighty river spirit seemed tragic, for it was locked away in shrines far from any water. Obviously, the spirit is invisible and resides in water, but the whole idea behind the spirit visiting and residing in shrines on land is precisely that these shrines are part of the compound of the healer who is believed to be the incarnation of the spirit. There were no road signs directing visitors to the shrines, which lie in an agricultural field far from the river, and all of them were secured with heavy padlocks. The shrines had never been in use and no healer lives there. Hence there is no point for a spirit to visit the shrines if the purpose is to convey messages through its human medium. The shrines are in practice not shrines, but merely empty structures without religious significance. Even the spirits have rejected them, despite all the efforts by the government and the funding provided by the dam's sponsors.

Consequently, notwithstanding all the rituals and signed and sealed certificates, the religious issues were not resolved. 'The EIB-CM underlines the importance of Jaja Bujagali as a recognised spiritual leader over the last decades. However, it ... recognises that the appeasement ceremonies as requested by Jaja Bujagali stand as an *unresolved issue* regarding the Bujagali Hydropower project.'[212] It is, however, highly unlikely there will be another appeasement ceremony. The dam is completed and the builders have closed their wallets and moved on. From their perspective, the case is closed, but that does not mean the spiritual issues are resolved. Although Jaja Bujagali was bypassed when Nfuudu allegedly moved the spirit several times, this does not mean from a cosmological standpoint that the Budhagaali spirit has accepted what happened. The falls have disappeared, but the spirit still resides there and is more active than before. In fact, it is raging because of what happened to Bujagali Falls.

Jaja Kagulu and the water spirits in context

High princely spirits or a clan spirit share one feature in common, concern for the wealth of their people. Although these spirits may be evoked in witchcraft, this does not compromise their overall cosmological role and importance in society. These spirits are genuinely there for their heirs and descendants. They may take the lives of commoners who undermine the social order or challenge divine authority, but this type of conduct is apparent in most religions. And among the Busoga, these spirits may manifest themselves in many places simultaneously and in various forms, and the understanding of these spirits and relations between them may differ from one healer's compound to another's.

In one, more local healer's compound, located outside Jinja at quite a distance from the lake and river, all the major spirits were present. In Busoga cosmology, Kintu is perceived as the father of all gods, and his wife is Nambi. There were four major water and lake spirits in this compound. In addition to Musoké the rain god, the spirit Kiwanuka was present. He also gave rain and his main responsibility was to bring wealth and blessings to the people. Although Mukasa is a Buganda spirit, he was also present, as was Lubaale, seen as the strongest water spirit of them all. Thus, there are multiple and overlapping perceptions of the spirits, and although the Buganda lake spirits strictly speaking belong to another cosmology and people, they may also be present among the Busoga. Moreover, Lubaale may be perceived as a water spirit by some, but not others.

Yet there are other spirits that have an indisputably central position in Busoga cosmology, one of them being Kagulu. Jaja Nabiryo Kagulu is the healer embodying this powerful spirit. She has one shrine in the town of Kamuli, but her main shrine is just beneath Kagulu Hill some 40 kilometres from Kamuli or 100 kilometres north of Jinja. Some see the Kagulu spirit as the most powerful of all Busoga spirits, and it is a manifestation of several water and hill spirits.

Jaja Kagulu is an elderly lady descended from the royal family and a daughter of Sir W.W.K. Nadiope (1911–77). Nadiope was elected by the Busoga council as king in 1949, initially for three years. In 1952 he was re-elected, and when the Owen Falls Dam was inaugurated he was one of the dignitaries receiving Queen Elizabeth. Jaja Kagulu is the third incarnation of the Kagulu spirit. The two held this position for about 50 years each before being succeeded. Currently, Jaja Kagulu has been the spirit's medium for 60 years. There may have been other earlier mediums, but she knows only of the above two. Thus she has been the longest serving Kagulu medium.

Kagulu Hill is a landmark in this part of Busoga land. Several hills or low mountains arise from the surrounding flats, and Kagulu is the largest of them, and thus the most powerful of all the spirits. However, there are many other spirits, such as Kapimpimi and Butimbinto. The Kagulu spirit is like a father to all the spirits. This is also of relevance to the origins of the royal family, and Jaja Kagulu is part of the royal lineage. From Jaja Kagulu's perspective, her spirit is the most powerful and the founding father of all. The Budhagaali spirit and the Itanda spirit are branches of the cosmology, but Kagulu is the epicentre. Jaja Itanda made a similar point: different spirits are like branches of a company, and the headquarters of all the spirits on water and land is Kagulu Hills, where all the spirits foregather. Jaja Bujagali also referred to Kagulu Hills and the spirit there.

As the founding father and foremost spirit, Kagulu empowers all the others. Even water spirits, including Budhagaali and Itanda, come to Kagulu Hill to receive strength. Although they perform best on or from water, they do receive powers from Kagulu, who distributes them like a father to his children. Even European spirits and the Christian God can come here and communicate with the Kagulu spirit. Jaja Kagulu also confirmed that not only was Jaja Bujagali the true Budhagaali incarnation, but the Budhagaali spirit is as powerful as it always has been, since it received power from Kagulu. This is despite the building of the dam, an improper and forceful transgression that violated the will of the spirit. However, the Budhagaali spirit did not stop dam construction as such, but avenges this sacrilege by creating havoc and killing people involved in the project.

Although the Kagulu spirit is now closely linked with the royal house, the spirit originally came from Ethiopia a long time ago. The Kagulu spirit's main role is to give power to others, and on its way from Ethiopia it dispensed power and gifts to those it passed. Eventually, the Kagulu spirit settled at this mountain. The reason for the spirit's decision to leave Ethiopia is unknown, but it was made by the spirit. One possible explanation is that there were very many spirits in Ethiopia, and so Kagulu decided to leave and start anew. However, what is important is the spirit's presence now, not its origins. The Kagulu spirit may also be viewed as Lubaale: the names differ but they refer to the same spirit, just as the Christian God and Allah are two sides of the same coin. There are different religions and names, but there is only one supreme god, Kagulu. Jaja Bujagali may argue otherwise – that it is the Budhagaali spirit that is most important. However, crucially, all spirits work in close cooperation and have relationships with other spirits, and this is a clue to understanding Busoga cosmology. Moreover, water spirits also move on land, and all spirits move around wherever they want. Still, the spirits have independent and particular 'homes', like Bujagali Falls.

The Budhagaali spirit

The Budhagaali spirit has always remained in the river or what is now the reservoir of the dam. All healers agree on this, apart from Nfuudu. Jaja Bujagali said

Figure 12. Caves in the Bujagali reservoir where sacrifices are made.

that the river spirit was never moved in the 2007 ritual. All the spirits have been there all the time. Not only is the Budhagaali spirit still living in the waters, so are innumerable other spirits and ancestors. Although most of the area has been submerged under the reservoir, some of the caves where other spirits reside and used for sacrifices still exist (Fig. 12). The lower caves have been flooded, but the uppermost are still in use. Sacrifices can be conducted any time during the year.

As Jaja Bujagali pointed out, ordinary people are not supposed to direct and command the spirits, rather the reverse is true. Consequently, the ritual initiated by the contractor and the government was improper. Similarly, when the medium is possessed, the spirit says what needs to be done and by whom – it is not for people to decide and say what the spirits should do.

Thus, throughout the whole process of construction, misconceptions about Busoga spirituality and water spirits in particular have led to many paradoxes, including the dam builder's attempts to remove the spirit and activist claims it would be destroyed. There is an inherent contradiction in the belief that a river god can be drowned by the construction of a dam. A river god can no more be destroyed than God or Jesus can be killed if a church is burnt down. Something as profane as a dam cannot destroy a god. In fact, it is impossible to drown a river spirit living in its own element. Moreover, residing in cascading waterfalls is a supreme manifestation of the powers of the spirit, but the waterfalls are *not* the spirit as such. Spirits are spirits and gods are gods, and their material embodiments and visualisations exist to allow humans to comprehend the divinities, but are not their essence. It is precisely because the Budhagaali spirit is a spirit that it is crucial for the whole of Busoga and not only for those living close to the river.

As a spirit living primarily in the waterfalls, the Budhagaali spirit was not pleased, to say the least, to see the falls disappear when they were flooded. Apart from Nfuudu, all agreed on this. And if spirits are benevolent when devotees adhere to divine principles, they may turn malevolent if people violate taboos and prescribed practices. Worse, when people not only neglect the divinities but also destroy the sacred places where they reside, apocalyptical wrath may ensue. This is what happened with the Budhagaali spirit, not unlike the deluge caused by God's wrath, but on a smaller scale within Busoga cosmology. The Budhagaali spirit became furious at the flooding of the falls, or so it was believed, and sought revenge by turning its wrath on all those involved in building the dam.

After the dam was completed and the hydropower station was activated, there were numerous technical problems and the project has not been as successful as expected, at least initially. From a technical and engineering point of view, such large projects often face teething problems. However, the functioning of the dam itself is a matter of dispute involving technology and cosmology. Does it work and deliver according to specifications or is this also a facade? The turbines have been temporarily out of use and the delivery capacity of the power station has not been optimal. According to Jaja Bujagali, the failures of the dam project are due to the river spirit. The spirit furiously attacks the hydropower station by every means and aims to harm the dam through its divine powers, since sacred places, including the rapids have been destroyed.

Even contractors and engineers allegedly claim that the problems they have faced with the dam's output are down to the Budhagaali spirit. Although officially the dam produces electricity at its full potential, this is not necessarily so from a religious point of view. Or more practically, is the electricity produced by hydropower alone or by hydropower topped up by generators? It was held by several that although the output is what the government says – 250 MW – the electricity produced by hydropower and the turbines is only a fraction of total output. The media reported technical problems and that the dam provided only 70 MW at the start – fuses blew and only one turbine was working. Thus, it was believed that the remaining 150 MW is produced by generators. Although the symptoms may be technical, the real cause of these problems was religious: it was the wrath of the Budhagaali spirit that created all the havoc at the dam. Contractors and engineers apparently come in secret to Jaja Bujagali asking him for the favour of ritually appeasing the spirit. Consequently, according to Jaja Bujagali, if everything is working, why are these people coming to him for advice and requesting him to please the spirit? Also, they are now coming to him after ignoring him for a long time as an imposter. Ashamed of their earlier actions and of ignoring him as the true incarnation, they now seek his advice. According to Jaja Bujagali, technocrats, engineers and World Bank representatives have visited him at his compound and asked if there is anything he can do to please the spirit so the dam can work properly. But he says there is nothing he can do when the spirit has been offended and this is the spirit's revenge: the spirit will only be placated when the rapids are returned, meaning that the dam is removed.

Moreover, several of the persons participating in the mock ritual transferring the spirit have died, and according to Jaja Bujagali, the healers who conducted the ritual are suffering from physical and mental illness. Throughout the process, he has maintained that if the dam is built and the falls destroyed, people will die. This is because of the spirit's wrath. Although as a healer he has the power to interact with the spirit, he stresses that he has done nothing and that what has happened is solely the will of the river spirit. He is only the medium, and the Budhagaali spirit does whatever it wants.

Why was Nfuudu chosen to conduct the ceremonies?

When all of the healers – not only Jaja Bujagali – say that the spirit is still there and has remained there all the time, why was Nfuudu allowed to perform the rituals seemingly relocating it? His behaviour, and indeed morality, was challenged by many, and one of the central healers of the land criticised him harshly and comprehensively. Not only was it impossible for him to relocate the spirit, since he was not its true incarnation, but also he had lied by claiming that he had successfully concluded the ritual. To make matters worse, he had even lied to President Museveni and the government by saying he had successfully removed the spirit from the falls. Thus, Nfuudu's behaviour was viewed with contempt. In traditional African religions, one does not move gods: it makes no sense and is impossible. It is gods that move people. Deceiving the president was not seen as a small matter by other healers, and had revealed the true character of Nfuudu. However in this case, it was not Museveni who had been deceived. Rather, Nfuudu's role in this ritual drama was orchestrated by the government all the way.

President Museveni is not a leader to take 'no' for an answer if he is determined to implement his plans. Despite the setbacks and termination of the first phase of the Bujagali project, the government was insistent the dam would be built. In the official view, NGOs and activists caused trouble and delay, and from an early stage it was evident that spiritual issues would have to be resolved in some way that would be seen as satisfactory by all parties, or at least those with an important say in the project. And it was also clear from the beginning that Jaja Bujagali would never allow the dam to be built. As the medium of the Budhagaali spirit, he would oppose it, and as long he was seen as legitimate, he would raise obstacles that might even lead to the termination of the second phase. The government, however, was not going to allow one healer and one spirit to block an almost $1 billion dam. The solution was to outmanoeuvre Jaja Bujagali, and Nfuudu was the answer.

Nfuudu was willing to perform the rituals believed to satisfy, appease and relocate the Budhagaali spirit. By conducting these rituals, there would be no hindrances to the completion of the dam. Thus, Nfuudu was brought in as the creation of a government aiming to resolve what it saw as a problem. As part of this strategy, it was necessary to undermine Jaja Bujagali's integrity and authority as the true incarnation of the Budhagaali spirit. The government participated in creating a split among the traditional leaders and healers attached to Budhagaali. Although money and support was officially given to both groups, including Jaja Bujagali, it was Nfuudu who enjoyed government backing. This strategy proved successful, and with the blessing of the Busoga kingdom Nfuudu acted as the main healer and officiant at the appeasement ceremonies to relocate the Budhagaali spirit.

In these murky waters, Nfuudu manoeuvred like a fish, gaining support from the kingdom and the chiefdom, although not without opposition. There were obviously external pressures being brought to bear on the kingdom to have Nfuudu accepted as the main ritual specialist, thereby enabling him to conduct the rituals, but there also seems to have been dissatisfaction within the kingdom with Jaja Bujagali as a healer and with his practices, despite his being the true Budhagaali. During the European Investment Bank's discussions with the kingdom in 2010, it became clear that were internal problems with the representativeness of both spiritual healers, but the kingdom was particularly displeased with Jaja Bujagali's numerous complaints against the dam contractor. Moreover, it seems the kingdom was particularly dissatisfied with the way Jaja Bujagali acted and lived: not only did he have many wives and eat meals in public, including dishes he was not allowed to consume, but his short hair was also noted.[213] In a study of traditional religion among the Busoga by the Cultural Research Centre in Jinja, these points were stressed. Jaja Bujagali is not allowed to eat fish or chicken; he is not allowed to share matoke with others except those possessed of the Nabamba spirit and confirmed to be so (and Jaja Bujagali is called the 'one who eats alone'); each new moon he must sleep alone without his wife for four days, and – he is not allowed to cut his hair.[214]

Some healers are bold and have cut their hair, like Nfuudu. Most others, however, have long hair and dreadlocks, like Jaja Bujagali and Jaja Itanda. Whether the healers have long hair or are shaven depends on the perceived ritual purity conveyed by their respective spirits. If the healer is free of debt to the spirit and ritually pure – in practice that the spirit is content with the ritual performances and obligations – the spirit may set the healer 'free.' The healer will still be the medium or the spirit's human incarnation, but the taboos and restrictions are no longer as strict. On the other hand, long hair on a healer is a sign that the spirit is not completely satisfied with the healer. This may have less to do with the healer's performance than with the whims and mood of the spirits: some spirits are more demanding and malevolent than others. Indeed, some spirits may demand so much that healers are practically incapable of fulfilling their wishes. These demands may include lavish rituals and extreme sacrifices, and much more. In cases where healers cannot fully comply, and it is the spirits who choose their medium and not the other way around, the spirits may choose not to let the healers 'free.'

Where healers have unfinished business with the spirits, cutting ones hair is a great sacrilege. Jaja Itanda said that if she cut her hair, her spirit would immediately strike her dead to the ground. Being a healer is thus both a blessing and also dangerous work, and one must obey divine laws and provide what the spirits demand. If cutting hair is certain death for the healer, there is one possible substitution for his or her own death. This is the ultimate human sacrifice, and in particular the sacrificing of a child to the river spirit. Among the Busoga there are many rumours and fears about these sacrifices. However, as noted there has been dissatisfaction with Jaja Bujagali for his having short hair. If it is correct that his long hair allowed Nfuudu to claim that he was the proper incarnation of the Budhagaali spirit, because he also had long hair, his apparent lack of long hair since, including his performance in the 2011 ritual, is striking. Thus, if such extreme rituals have been conducted recently, and there are widespread rumours that they have, but no evidence, there is no single and obvious candidate. Importantly, these dark sides of spirituality and witchcraft are not meant to be public but to remain shrouded in mystery. People may believe in this reality, but for obvious reasons no outsider may participate in them, since then they would become witnesses to murder and hence accomplices.

Religious aftermath

Despite the fierce fighting among the healers and the kingdom's support for Nfuudu, thereby leaving Jaja Bujagali out in the cold, things have changed. The 2011 ritual was the last straw and changed the kingdom's attitude towards Nfuudu, simply because it was not a ritual, and afterwards he became *persona non grata*. Today, he is no longer the chief healer in the kingdom and a new one has been appointed with no affiliation with the Bujagali Falls or the Budhagaali spirit.

It is now generally acknowledged that the 2011 ritual was a staged and pretended ritual – a charade. Nfuudu was not possessed when the rite was performed, and the ritual did not convince participants, with the consequence that Nfuudu was seen as a charlatan and Jaja Bujagali as having been right all along.

The improper ritual performances convinced people that they had been deceived and the kingdom realised it had been fooled by Nfuudu. It was impossible to transfer the spirit in the first place, and consequently impossible to transfer it back or to the Namizi shrines. A healer can only be possessed by one spirit and by one alone. Anyone claiming to be possessed by another or other spirits is lying, a chief said. He is only bragging and full of himself, but multiple possession is not possible in practice. This is, however, only partly true, since other healers claim to be possessed by several spirits.

Although it was claimed that Nfuudu moved the spirit in 2011, this claim was open to diverse interpretations. One is, of course, Nfuudu's version. Another is that this is proof that if he relocated the spirit it was not, by definition, in his possession – he was not the owner of the spirit or its incarnation. He could have been the custodian of the spirit for a short time, but he was not the proper Jaja Bujagali. Jaja Bujagali was the real and only medium of the Budhagaali spirit. Nfuudu was an imposter, and employing him as the healer for the rituals had fatal implications. The Budhagaali spirit was not pleased with what had happened. And when the kingdom realised that the 2011 ritual was a charade, the implication was that the rituals conducted in 2001 and 2007 were also staged and fake.

Chief Ntembe of the Busoga kingdom, who made the agreement with the contractor and the government allowing Nfuudu to perform the ritual transfer of the Budhagaali spirit in 2007, died 18 May 2009. He was a very old man and his death would sooner or later have been inevitable, but it was understood to be in consequence of Budhagaali's wrath. In the kingdom, it was generally acknowledged that he had been deceived and forced to accept the agreement allowing Nfuddu to perform the rituals paving the way for the dam, but that is no excuse in a divine perspective. Not only did his actions bring divine wrath, but they also had direct consequences in that the chief was harshly punished and killed by the Budhagaali spirit. The chief did not die of natural causes. This was evidently confirmed by his ancestors, who told of fatal demons in his head that were the work of the furious Budhagaali spirit. The old chief was killed by the spirit for not adhering to the cosmic rules.

Waguma Yasin succeeded Chief Ntembe, his father. A central feature of Busoga culture is reconciliation.[215] Jaja Bujagali has a reputation of being intensely hostile to his enemies, but also of sincerely loving and taking care of his devotees and friends. During the summer of 2013 Chief Waguma met Jaja Bujagali. Despite the bitterness and conflicts aroused by the dam project, and in particular the kingdom's choice of Nfuudu as its representative and not Jaja Bujagali, the custodian and embodiment of the spirit, Chief Ntembe was reconciled with Jaja Bujagali. Not only had money made everything worse and the wrong decision brought on the wrath of the Budhagaali spirit and the death of his father, but the experience of being deceived in what mattered most, namely religion, included the whole kingdom, with serious implications. Religion matters and the Budhagaali spirit is still the most important deity within the chiefdom, apart from the Christian and Muslim gods. Jaja Bujagali is its medium and Nfuudu is a charlatan. Thus, the cosmological order has been restored, not to where it was before the dam project started, because much water has flown in the river since then causing many upheavals, but to the extent that Jaja Bujagali is the undisputed medium of the Budhagaali spirit (Fig. 13). And that spirit is one of the most important in the Busoga cosmology. Moreover, Jaja Bujagali has regained his central position in the kingdom as one of its main ritual specialists. On Saturday, 13 September 2014, William Gabula Nadiope was enthroned as the new King (Kyabazinga) of Busoga. President Museveni was in attendance as were many other dignitaries. In the ceremony, Basoga traditional culture also featured. Jaja Bujagali was instrumental in the impressive rituals associated with the 'traditional' events, which included the spirit world and the intangible parts of Basoga heritage.

Revenge of the spirits

As noted above, among many of those believing in traditional cosmology it was perceived that the Budhagaali spirit was infuriated by the misconduct associated with the dam, and avenged himself by killing those who allowed the project to proceed. Although Jaja Bujagali had consistently predicted these consequences, nobody listened to him or did not care. From a religious understanding of the wrath of the spirits, it seems that history may repeat itself, probably at Kalagala or Itanda Falls, and maybe also at Murchison Falls, where preliminary studies have indicated a hydropower potential of 642 MW.[216] If and when this dam is built, is uncertain.

The water spirits at Murchison Falls have played a prominent role in Uganda's history, with serious repercussions up to today, even if one does not believe in spirits and the indigenous cosmology (Fig. 14). The Christian God sent the spirit Lakwena to Uganda on 2 January 1985. On 15 May of that year, Lakwena violently possessed Alice Auma, an Acholi in northern Uganda, Lakwena meaning 'messenger' in the Acholi language. The spirit ordered her to a place called Wang Jok by Murchison Falls, where she stayed for 40 days. It is unclear to what extent this move sprang from Alice's conviction or her father's persuasiveness, but this event was important to her later life story.[217] On 29 May, the spirit held court on the water, and at its command the water in the falls stopped flowing. Lakwena asked: 'Water, I am coming to ask you about the sins and the bloodshed in this world,' whereupon the spirit in the falls answered: 'The people with two legs kill their brothers and throw the bodies into the water … I fight against the sinners,

Figure 13. Jaja Bujagali.

Figure 14. Murchison Falls.

for they are the ones to blame for the bloodshed. Go and fight against the sinners, because they throw their brothers in the water.'[218]

The spirit convinced Alice that the reason the country was being torn apart by civil war was because of the pervasiveness of witchcraft at all levels, involving personal greed and power over one's opponents. The battle was seen as a preliminary Last Judgment to purify the world of evil. Death and being killed did not undermine her belief system, since they were viewed as justifiable punishment for the sins. On 6 August 1986, Lakwena instructed Alice to give up her local medicine practice and to become a military commander. A report the Holy Spirit Movement provided to missionaries in June 1987 stated: 'The good Lord who sent Lakwena decided to change his work from that of a doctor to that of a military commander for one simple reason: it is useless to cure a man today only that he be killed the next day. So it is an obligation on his part to stop the bloodshed before continuing his work as a doctor.'[219]

Alice formed the Holy Spirit Movement with the aim of ending the civil war and cleansing Uganda of sin, and in August 1986 she formed an army called the Holy Spirit Mobile Forces. And she was by no means alone in this project. Some 7,000–10,000 Holy Spirit Soldiers joined the ranks along with 140,000 spirits, and even bees, snakes, rivers, stones and mountains. Although the soldiers were armed with stones and guns, water was also prominent as a weapon and source of protection in the fight, for the spirit had said: 'There is nothing greater than water,' and 'Whatever it is, it will be washed away by water!'[220]

Lakwena was not an indigenous spirit in the traditional sense. According to one soldier: 'Lakwena is a holy spirit. Now being a spirit he is not visible. Nobody has seen … Lakwena and we should not expect to see him anyway. Being a spirit he has no relatives on earth. He speaks 74 languages including Latin … When the Holy Spirit is addressed he should be called "Sir".'[221] Lakwena was the supreme commander, but there were other spirits as well. All of the spirits in the Holy Spirit Movement were in fact 'foreign' or 'alien' spirits in the local cosmology. Lakwena was the spirit of an Italian captain who had died during or after the Second World War near Murchison Falls. Another account has him drowning in the Nile during the First World War. Then there was the Wrong Element spirit, from the United States; the Franko spirit from Zaire; the Ching Poh spirit from China or Korea; and a number of alien Christian and Arab spirits.[222]

Although the movement enjoyed initial success, it was defeated by the National Resistance Army led by Yoweri Museveni. Thus, during the civil war Museveni fought against movements prompted by a river spirit that urged the Acholi to take up arms. It is perhaps no wonder that Museveni would not allow a river spirit at Bujagali Falls to halt his plans for a hydroelectric dam: he had fought against stronger water spirits before with more lethal consequences. Given the bloody civil war and its spiritual roots in the water spirit at Murchison Falls, one should not hold one's breath that the preservation of indigenous religion and water spirits will be a high priority if a dam is built at this place.

Although Alice Lakwena set out to combat the evil of witchcraft, witchcraft accusations also became her fate. Despite her being a prophet, there was distrust in the ranks. When the Holy Spirit Mobile Forces reached Busoga en route to Kampala, Lakwena instructed the soldiers not to use the bridge to cross the Nile, but to walk on the water like Jesus. Only 300 were allegedly pure enough for this, and sinners would remain on the shore. This caused disbelief among the soldiers, some of whom deserted. After the defeat by the National Resistance Army, many former Holy Spirit soldiers lost faith in Lakwena. Some claimed she was a witch; others that she was a prostitute with AIDS who did not want to die alone, and had recruited the army so that as many as possible would die with her. In Kampala, other rumours flourished among opponents of the movement: Museveni had recruited another witchdoctor to give National Resistance Army soldiers strong medicines to enable them to defeat their Holy Spirit adversaries.[223] Whatever its truth, in a same vein Nfuudu is allegedly providing medicines to soldiers fighting against rebel groups on the borders of the Congo, including the Lord's Resistance Army.

And it was precisely on the ruins of the Holy Spirit Movement that Joseph Kony built the Lord's Resistance Army, one of the most brutal armed movements in modern history. Kony claimed to be a cousin of Alice. He too became possessed by a particular spirit, named Juma Oris, who had been a minister under Idi Amin. This spirit became Kony's chief spirit and chairman, and ordered Kony to liberate humanity from disease and suffering. It also held that healing was meaningless when those healed were killed. Kony did not inherit Alice's spirits, but introduced a range of completely new and unknown ones.[224] In 2005 the International Criminal Court issued an arrest warrant for Kony, charging him with 12 counts of crimes against humanity and 21 counts of war crimes. The ideology with which he was associated had started 20 years earlier with the prophecies of one healer and on the orders of the water spirits at Murchison Falls.

Thus, one sees a complex interplay between spirits, actors and factors. Lakwena was not a river spirit, but he guided Alice to Murchison Falls were she received her instructions from other spirits. Thus, Lakwena, a foreign and Christian spirit, cooperated closely with local, indigenous spirits. Although Lakwena as an independent spirit demanded that Alice took up arms, it was the miseries explained by the water spirits that prompted the movement. Kony too used water in his possession, but he included a wide range of spirits unrelated to the Nile in his cult. However, through the Holy Spirit Movement, the Lord's Resistance Army also originated in part in the Nile spirits in Murchison Falls.

From a cultural heritage perspective, Lakwena presents more challenges and paradoxes than easy answers for the preservation of indigenous religion and water spirits in an era of dam building. All these examples, including the now offended Budhagaali spirit who wreaks havoc on society, clearly show that spirits are believed to have independent powers and agency, and are not necessarily good or bad, but both. Regardless of whether the spirits objectively exist or not, their presence is a matter of faith and belief, and Christians obviously reject it. 'The essence of religion is not even our concern, as we make it our task to study the conditions and effects of a particular type of social behavior,'[225] Max Weber said. Thus, the fact that some believe that spirits exist may have unexpected and at times devastating consequences, as the history of Alice Lakwena exemplifies. What will happen if the Itanda spirit is offended, nobody knows. And given the history of spirits at Murchison Falls, the religious interpretation of the ravaging spirits' needs and concerns may have serious consequences in the secular sphere. These may be felt far away from the healers who become possessed, for the spirits may work together and water spirits have to be seen in relation to all the other spirits in a given cosmology.

Future dams, people and their spirits
So what to do about the water spirits when dams are built? The answers are provided by the healers themselves, and they do not involve relocating the spirits, as Nfuudu claimed. The idea of transferring a spirit, in this case a water spirit, is in itself a paradox. On one hand, the healer is believed to take good care of the

spirits, but the spirits may walk on land or move about in whatever form and shape they wish. Thus, it is not possible to transfer a spirit against its wishes, for it can simply move back to where it was. On the other hand, this possibility also underscores the dual aspect of the spirits. It is the very waters and power of the falls or source that express the powers of the spirits, but the spirits are at the same time omnipresent and can be far from the physical manifestation of their forces. In another context, Obeyesekere noted, 'not only are they [gods] present in a particular community, but they may be present if invoked in other communities and shrines at the same time. They must be in this place, and that, in the then and the now. They therefore obviously cannot be present *in person*; rather, they are there in *essence*.'[226] Thus, the spirits are both the water and its force while at the same time invisible and above and apart from the very powers that define them. Consequently, the transfer of a water spirit like Budhagaali is a contradiction: it is no more possible for a healer to transfer a Busoga spirit than it is possible for a pope to transfer the Holy Spirit in Christianity.

However, as all the healers stress, it is possible to appease the spirits, as long as the right healer conducts the appropriate rituals. The spirits are not necessarily against dams and development, but they want the cosmological hierarchy to be acknowledged and respected. When that has been achieved, they may allow the dams to proceed. However, who the proper healers are, and what the necessary rituals can be, as we have seen, are contested by other healers claiming to be the mediums for the very same spirits. Resolving this contest requires in-depth knowledge of ongoing processes at each place. Given the above discussion regarding the Bujagali Dam, the social and environmental assessment studies underestimated the spiritual complexity. That said, the builder seems to have done its best given the existing knowledge, and also given the Busoga kingdom's support for Nfuudu at the time.

This episode also raises questions about international NGOs' and campaigners' unanimous support for protecting indigenous religion. Apart from the fact that cultural and religious processes are constantly changing, this firm support for a cosmology that also includes witchcraft and, according to rumours, child sacrifice seems to have been too hasty and not based on full knowledge of all the religious practices in question. From a broader societal perspective, it also points to the fact that how these spirits are understood may have far-reaching consequences. How the spirits will react, or more precisely how believers will react as they interpret the instructions of their spirits, is an open question. A dam at Murchison Falls may have serious unintended consequences if the spirits are believed to be annoyed. Thus, whether the spirits are good, bad or both, is in the eyes of the beholder. In any event, the spirits are not against humans and their welfare and betterment, rather the contrary: the Nile is wealth in both the profane and secular spheres.

The complexity of the interrelationships among the spirits – both on land and water – is key to understanding the impacts and consequences of, for instance, the Bujagali Dam, which flooded the sacred falls where the Budhagaali spirit resided (although it still resides in the reservoir). The spirits in Owen Falls were also flooded, and so in all likelihood will the spirits of Itanda and of other places along the Nile be. The beautiful scenery admired by Stanley and Churchill will disappear, but both men were, as Museveni is, willing to sacrifice beauty for benefits in the form of hydropower and industrialisation. Although political processes are not inevitable, it seems probable that all these falls will disappear in the near future. Whether one is for or against dams, and there are arguments favouring both positions, one has to acknowledge the autonomy of independent states in choosing their development paths. Certainly, Uganda needs energy for development, and dams cause controversies.

Still, after critically examining what happened and still happens at Bujagali Falls, little of what extreme anti-dam activists initially claimed still holds water. The claim that the Bujagali Dam would lead to the cultural death of 2.5 million Busoga and possible ethnic cleansing was pure fancy. The spirits are different: they will continue to exist forever, and nothing as mundane as a dam can, believers hold, challenge the existence of the spirits and gods. In this regard, Busoga spirits are like the Christian God: neither God, Jesus, the Holy Spirit nor the Virgin Mary would be extinguished by a dam. Or, in the case of Islam, if Mecca were destroyed, Allah would not be killed, but the destruction would have divine consequences and believers will behave accordingly.

So it is with the Busoga spirits, although they do not work in the same way and operate in a more local context. The waterfalls are the true testament of the spirits' powers, and when the falls are gone, their powers diminish, or so it is believed. Ontologically and cosmologically, the spirits are as powerful as ever, but there is another important fact: despite the allegedly autonomous powers of the spirits, even they have to prove their strength. This can be through the outcomes of witchcraft, but also through the force of the falls. The disappearance of the falls attests to the disappearing powers of the gods and healers, or so it was claimed.

Some have already started to question Jaja Bujagali's powers now the falls have disappeared. What is certain, however, is that the spirits still exist and will continue to do so: spirits are spirits, and humans are humans. And humans die, but not spirits, and after death humans become part of the ancestral world of the spirits. What is also certain is that people and believers choose the healers they perceive to be the most powerful ones in engaging with the ancestors and spirits, or witchcraft. Even if individual spirits lose some powers because the waterfalls are gone, as some commoners believe, Busoga cosmology still exists. With thousands of healers engaging with the various spirits, there are always other healers who can provide and guarantee divine intervention – and deliver outcomes. Thus, rather than dams along the Nile in Uganda causing cultural death or the extinction of beliefs, it is more likely they cause a transfer of power and alteration of hierarchies among healers and their spirits.

In any event, dam construction will continue to fuel heated debate, in which there is no middle ground, in the sense that the dam is either built or not. However, the culturally specific contexts where dams are built are, by definition, not universal and a global discourse either in favour of or against dams is not necessarily the most fruitful. Dams alter specific cultural worldviews and cosmologies, which may vary greatly from place to place. At the outset, more thorough assessment studies may be required that document the traditions and practices at stake. Even so, given the religious complexity of a dam site like Bujagali Falls, probably no intervention will satisfy all believers and healers, despite all the efforts to appease and please devotees and divinities. Not only are there internal hierarchies and disagreements among healers and within a cosmology, traditions are also always changing and adapting to new realities. Although Jaja Bujagali today says that if the appropriate rituals were conducted by him, the dam could have been built, before the dam was constructed he insisted that the spirit would never allow the dam. From this perspective, dams are not good or bad, but both or something in the middle. Moreover, traditional or indigenous culture and religion are not always unambiguously good, so that protecting them in pristine form may not always be desirable, even if it were possible. Cultures change all the time, but one thing this study has shown is that the powers of the water continue and both unite and transcend social, political and cosmological spheres. The powers of Bujagali Falls are not tamed even though they are dammed, and the forces continue in this and otherworldly realms.

References

African Development Bank 2009. African Development Bank management plan in response to the independent review panel's report on the Bujagali hydropower and interconnection projects. Unpublished.

Anderson, K. 2009. Review: What NGO Accountability Means: And Does Not Mean. NGO Accountability: Politics, principles and Innovations, by Lisa Jordan and Peter van Tuijl (eds.). *American Journal of International Law* 103(1): 170–8.

Baker, S.W. 1869. *Albert N'yanza. Great Basin of the Nile, and Exploration of the Nile Sources.* Lippincott. Philadelphia.

Behrend, H. 1999. *Alice Lakwena & the Holy Spirits.* James Currey. Oxford.

Beke, C.T. 1847. On the Nile and Its Tributaries. *Journal of the Royal Geographical Society of London,* Vol. 17: 1–84.

Borde, B.H. and Y. Charafi. 2013. An inconvenient truth. *Private Sector & Development. Proparco's Magazine,* No. 18 (November): 9–12.

Bosshard, P. 2010. China Dams the World. *World Policy Journal* 26(4) (Winter 2009/2010): 43–51.

Briscoe, J. 2010. Viewpoint – Overreach and Response: The Politics of the WCD and its Aftermath. *Water Alternatives* 3(2): 399–415.

Briscoe, J. 2011. Invited opinion interview: Two decades at the centre of world water policy. Interview with John Briscoe by the *Water Policy* Editor-in-Chief. *Water Policy* 13: 147–60.

Burnside. 2006. Bujagali Hydropower Project. Social and Environmental Assessment Report. Executive Summary. R.J. Burnside International. Canada.

Burton, R.F. and J.H. Speke. 1859. Explorations in Eastern Africa. *Proceedings of the Royal Geographical Society in London* 3(6) (1858–1859): 348–85.

Byerley, A. 2005. *Becoming Jinja. The Production of Space and Making of Place in and African Industrial Town.* Department of Human Geography, Stockholm University.

Churchill, W.S. 1909. *My African Journey.* William Briggs. Toronto.

Cohen, D.W. 1986. *Towards a reconstructed past: Historical texts from Busoga, Uganda.* Oxford University Press. Oxford.

Collins, R.O. 2002. *The Nile.* Yale University Press. New Haven CT.

Conca, K. 2006. *Governing water: contentious transnational politics and global institution building.* MIT Press. Cambridge MA.

EIB-CM 2012. *European Investment Bank Complaints Mechanism (EIB-CM). Bujagali Hydroelectric Project, Jinja, Uganda. Complaint SG/E/2009/09.* Conclusion Report. 30 August.

Environmental Defence, Friends of the Earth and International Rivers Network 2003. *Gambling with People's Lives. What the World Bank's New "High-Risk/High-Reward" Strategy Means for the Poor and the Environment.* Environmental Defence, Friends of the Earth & International Rivers Network, Washington, Amsterdam and Berkeley.

Fallers, L.A. 1969. *Law without Precedent. Legal Ideas in Action in the Courts of Colonial Busoga*. University of Chicago Press. Chicago.

Findlay, A.G. 1867. On Dr. Livingstone's Last Journey, and the Probable Ultimate Sources of the Nile. *Journal of the Royal Geographical Society of London* 37: 192–212.

Garstin, W. 1909. Fifty Years of Nile Exploration, and Some of Its Results. *Geographical Journal* 33(2): 117–47.

Gonza, R.K. et al. 2001. *Reconciliation among the Busoga*. Cultural Research Centre, Jinja.

Gonza, R.K. et al. 2002. *Traditional Religion and Clans Among the Busoga*. Vol. 1. Cultural Research Centre. Jinja.

Heien, K.H. 2007. *Local Livelihoods and the Bujagali Hydro-Power Dam, Uganda*. MA thesis, Faculty of Economics and Social Sciences, Agder University College. Kristiansand.

Hoag, H.J. 2013. *Developing the Rivers of East and West Africa. An Environmetal History*. Bloomsbury. London.

Hoyle, B.S. 1963. The Economic Expansion of Jinja, Uganda. *Geographical Review* 53(3): 377–88.

International Development Association .Report No. IDA/R2008–0296. International Bank for Reconstruction and Development. International Development Association. Management Report and Recommendation in Response to the Inspection Panel Investigation Report.

Joesten, J. 1960. Nasser's Daring Dream: The Aswan High Dam. *World Today* 16(2): 55–63.

Johnston, H. 1903. *The Nile Quest*. Lawrence and Bullen. London.

Lubwama, N.Z. 2012. The meanings of heritage: Practices, spaces and sites in the Busoga Kingdom, Uganda in the twentieth first century. Thesis for Magister History, Department of History, University of Western Cape.

Lugard, F.J.D. 1893. *The Rise of Our East African Empire: Early Efforts in Nyasaland and Uganda*. Vol. 2. Blackwood and Sons. London.

Majot, J. 2003. Comment in *World Rivers Review* 18(5): 1–16.

Mallaby, S. 2004. NGOs: Fighting Poverty, Hurting the Poor. *Foreign Policy*, No. 114 (Sep.– Oct): 50–8.

McCully, P. 2001. The Use of a Trilateral Network: An Activist's Perspective on the Formation of the World Commission on Dams. *American University International Law Review* 16, Issue 6: 1453–75.

Mill, J. S. 1865. *Principles of political economy, II*. London.

Moorehead, A. 2000[1960]. *The White Nile*. Perennial. New York.

Murchison, R. 1863. Address to the Royal Geographical Society. *Journal of the Royal Geographical Society of London*, Vol. 33, No: cxiii–cxcii.

Mutambi, B. 2012. Sustainable energy for all initiative. How prepared is Uganda? *Electricity Regulatory Authority.* Newsletter, Issue 7 (December): 4–6.

Mutyaba, V. 2012. Uganda at 50 and the Electricity sub-sector 58: What does the future hold? *Electricity Regulatory Authority.* Newsletter, Issue 7 (December): 10–12.

Nayenga, P.F.B. 1979. Chiefs and 'Land Question' in Busoga District, Uganda, 1895–1936. *International Journal of African Historical Studies* 12(2): 183–209.

Nwauwa, A.O. 2000. The Europeans in Africa: Prelude to Colonialism. In Falola, T. (ed.) Africa. Vol. 2. African Cultures and Societies before 1888: 303–18. Carolina Academic Press. Durham. *Africa. Vol 2. African Cultures and Societies before 1888.*

Obeyesekere, G. 1984. *The Cult of the Goddess Pattini.* University of Chicago Press. Chicago.

Oestigaard, T. 2012. Water Scarcity and Food Security along the Nile: Politics, population increase and climate change. *Current African Issues* No. 49. Nordic Africa Institute. Uppsala.

Oweyegha-Afunaduula, F.C. 2003. Huge dams as corporate crime and terrorism. *NAPE LOBBY*, November, 6th ed., Vol. 2(4): 21.

Probe Alert. 2001. *World Bank Set to Aid Multinational Power Company, Not Africa's Poor.* September. Probe Alert. Toronto.

Speke, J.H. 1863. *Journal of the Discovery of the Source of the White Nile.* Blackwood and Sons. Edinburgh and London.

Stanley, H.M. 1878. A Geographical Sketch of the Nile and Livingstone (Congo) Basins. *Proceedings of the Royal Geographical Society of London* 22(6) (1877–1878): 382–410.

Stanley, H.M. 1898. *Africa: Its Partition and Its Future.* Dodd, Mead and Company. New York.

Tangri, R. and A.M. Mwenda. 2013. *The Politics of Elite Corruption in Africa. Uganda in Comparative African Perspective.* Routledge, London.

Thomas, H.B. and I.R. Dale. 1953. Uganda Place Names: Some European Eponyms. *The Uganda Journal* 17(2): 101–23.

Tvedt, T. 2004. *The River Nile in the Age of the British. Political Ecology and the Quest for Economic Power.* I.B.Tauris. London.

Tvedt, T. 2010. Why England and not China and India? Water systems and the history of the industrial revolution. *Journal of Global History* 5: 29–50.

Tvedt, T. 2011. Hydrology and Empire: The Nile, Water Imperialism and the Partition of Africa. *Journal of Imperial and Commonwealth History* 39(2): 173–94.

Tvedt, T. 2012. *Nilen – historiens elv.* Aschehoug. Oslo.

UN World Water Development Report 2014. *Water and Energy.* Vol. 1. Unesco. Paris.

Waterbury, J. 1979. *Hydropolitics of the Nile Valley.* Syracuse University Press. New York.

Waterbury, J. 2002. *Nile Basin. National Determinants of Collective Action.* Yale University Press. New Haven CT.

World Bank IP 2002. World Bank. 2002. *Report No. 23988. The Inspection Panel. Investigation Report. UGANDA: Third Power Project (Credit No. 2268-UG), Forth Power Project (Credit No. 3545-UG) and Bujagali Hydropower Project (PRG No. B 003-UG), May 23, 2002.* World Bank. Washington.

World Bank. 2003. *Accountability at the World Bank: The Inspection Panel 10 Years On.* World Bank. Washington.

World Bank IP 2008. World Bank. 2008. World Bank 2008. *Report No. 44977-UG. The Inspection Panel. Investigation Report. Uganda: Private Power Generation (Bujagali) Project (Guarantee No. B0130-UG). August 29, 2008.* World Bank. Washington.

World Commission on Dams. 2000. *Dams and Development. A New Framework for Decision-Making. The Report of the World Commission on Dams.* Earthscan. London.

Weber, M. 1964. *The Sociology of Religion.* Beacon Press. Boston.

Wilson, G. 1967. Owen Falls. Electricity in a Developing Country. *East African Studies* 27. East African Publishing House. Nairobi.

Endnotes

1. There are different years in the literature: 1613, 1615 and 1618, see Johnston, H. 1903. *The Nile Quest*. Lawrence and Bullen. London, p. 51.

2. Beke, C.T. 1847. On the Nile and Its Tributaries. *Journal of the Royal Geographical Society of London*, Vol. 17: 1–84, p. 2.

3. Murchison, R. 1863. Address to the Royal Geographical Society. *Journal of the Royal Geographical Society of London*, Vol. 33, No: cxiii–cxcii, p. clxxiv.

4. Speke, J. H. 1863. *Journal of the Discovery of the Source of the White Nile*. Blackwood and Sons. Edinburgh and London, p. 495.

5. Speke 1863: 466–467.

6. Speke 1863: 469.

7. Moorehead, A. 2000[1960]. *The White Nile*. Perennial. New York, p. 142–143.

8. Burton, R.F. and J.H. Speke. 1859. Explorations in Eastern Africa. *Proceedings of the Royal Geographical Society in London* 3(6) (1858–1859): 348–85, p. 355.

9. Baker, S.W. 1869. *Albert N'yanza. Great Basin of the Nile, and Exploration of the Nile Sources*. Lippincott, Philadelphia, p. xxi.

10. Mill, J. S. 1865. *Principles of political economy, II*. London. p. 122.

11. Baker 1869: xxii.

12. Tvedt, T. 2012. Nilen – historiens elv. Aschehoug. Oslo.

13. Baker 1869: xxi, my emphasis.

14. Garstin, W. 1909. Fifty Years of Nile Exploration, and Some of Its Results. *The Geographical Journal*, Vol. 33, No. 2: 117–147, p. 117.

15. Garstin 1909: 118.

16. Tvedt, T. 2004. *The River Nile in the Age of the British. Political Ecology and the Quest for Economic Power*. I.B.Tauris. London; Tvedt, T. 2010. Why England and not China and India? Water systems and the history of the industrial revolution. *Journal of Global History* (2010) 5: 29–50; Tvedt, T. 2011. Hydrology and empire: The Nile, Water Imperialism and the Partition of Africa. *The Journal of Imperial and Commonwealth History*, Vol. 39, No. 2: 173–194.

17. Tvedt 2011: 175.

18. Tvedt 2011: 177.

19. op. cit. Tvedt 2011: 176.

20. Tvedt 2004: 22.

21. op. cit. Tvedt 2011: 182

22. op. cit. Tvedt 2011: 182

23. op. cit. Tvedt 2011: 187.

24. Tvedt 2010.

25. Churchill, W. S. 1909. *My African Journey*. William Briggs. Toronto, p. 24.

26. Churchill 1909: 61.

27. Churchill 1909: 62–63.

28. Churchill 1909: 123.

29. Churchill 1909: 209, 211.

30. Churchill 1909: 197.

31. Lugard, F. J. D. 1893. *The Rise of Our East African Empire: Early Efforts in Nyasaland and Uganda. Vol. 2*. Blackwood and Sons. London, p. 3.

32. Stanley, H. M. 1898. *Africa: Its Partition and Its Future*. Dodd, Mead and Company. New York, p. 63–64.

33. Churchill 1909: 119.

34. Churchill 1909: 120.

35. Churchill 1909: 121, 123.

36. Stanley, H. M. 1878. A Geographical Sketch of the Nile and Livingstone (Congo) Basins. *Proceedings of the Royal Geographical Society of London*, Vol. 22, No. 6 (1877 – 1878): 382–410, p. 390.

37. Stanley 1878: 391.

38. op. cit. Stanley 1878: 409.

39. Fallers, L. A. 1969. *Law without Precedent. Legal Ideas in Action in the Courts of Colonial Busoga.* The University of Chicago Press. Chicago, p. 41.

40. Heien, K. H. 2007. *Local Livelihoods and the Bujagali Hydro-Power Dam, Uganda. MA-thesis. Faculty of Economics and Social Sciences.* Agder University College. Kristiansand.

41. Byerley, A. 2005. *Becoming Jinja. The Production of Space and Making of Place in and African Industrial Town.* Department of Human Geography. Stockholm University. Stockholm, p. 1.

42. Hoag, H. J. 2013. *Developing the Rivers of East and West Africa. An Environmetal History.* Bloomsbury. London, p. 177.

43. http://uegcl.com/kiira-Power.html, http://uegcl.com/nalubaale-power.html (accessed 21 August 2014)

44. World Bank. 2003. *Accountability at the World Bank: The Inspection Panel 10 Years On.* World Bank, Washington, p. 84.

45. Nayenga, P. F. B. 1979. Chiefs and 'Land Question' in Busoga District, Uganda, 1895–1936. *The International Journal of African Historical Studies*, Vol. 12, No. 2: 183–209, p.189.

46. Hoyle, B. S. 1963. The Economic Expansion of Jinja, Uganda. *Geographical Review*, Vol. 53, No. 3: 377–388, p. 379.

47. Hoyle 1963: 388.

48. Churchill 1909: 133.

49. Tvedt 2004: 217.

50. Waterbury, J. 1979. *Hydropolitics of the Nile Valley.* Syracuse University Press. New York, p. 89.

51. Wilson, G. 1967. Owen Falls. Electricity in a Developing Country. *East African Studies 27.* East African Publishing House. Nairobi, p. 5–7.

52. Tvedt 2004: 224.

53. Collins, R. O. 2002. *The Nile.* Yale University Press. New Haven, p. 158.

54. Waterbury, J. 2002. *Nile Basin. National Determinants of Collective Action.* Yale University Press. New Haven, p. 160.

55. Tvedt 2004: 223.

56. Tvedt 2004: 308–309.

57. Byerley 2005: 264.

58. Thomas, H.B. and I.R. Dale. 1953. Uganda Place Names: Some European Eponyms. *The Uganda Journal* 17(2): 101–23, p. 117.

59. Probe Alert September 2001. *World Bank Set to Aid Multinational Power Company, Not Africa's Poor.* Probe Alert. Toronto.

60. African Development Bank 2009. *African Development Bank management plan in response to the independent review panel's report on the Bujagali hydropower and interconnection projects.* Unpublished, p. 2.

61. *Electricty Regulatory Authority Newsletter*, Issue 7, December 2012, p. 16.

62. EIB-CM 2012. *European Investment Bank Complaints Mechanism (EIB-CM). Bujagali Hydroelectric Project, Jinja, Uganda. Complaint SG/E/2009/09.* Conclusion Report. 30 August 2012, p. 28.

63. WB IP 2008. World Bank 2008. *Report No. 44977-UG. The Inspection Panel. Investigation Report. Uganda: Private Power Generation (Bujagali) Project (Guarantee No. B0130-UG). August 29, 2008.* World Bank. Washington, p. 29.

64. Burnside 2006. Bujagali Hydropower Project. Social and Environmental Assessment Report. Executive Summary. R. J. Burnside International Limited. Canada, p. 1, 27.

65. Burnside 2006: 40.

66. Burnside 2006: 17–18.

67. Burnside 2006: 28.

68. Burnside 2006: 33.

69. Burnside 2006: 42.

70. WB IP 2008: 135–136.

71. International Development Association .Report No. IDA/R2008–0296. International Bank for Reconstruction and Development. International Development Association. Management Report and Recommendation in Response to the Inspection Panel Investigation Report, p. 25.

72. Joesten, J. 1960. Nasser's Daring Dream: The Aswan High Dam. *World Today*, Vol. 16, No. 2: 55–63, p. 59.

73. Briscoe, J. 2010. Viewpoint – Overreach and Response: The Politics of the WCD and its Aftermath. *Water Alternatives* 3(2): 399–415, p. 401–402.

74. http://www.internationalrivers.org/files/attached-files/the_world_banks_big_dam_legacy.pdf (accessed 13 February 2014)

75. Environmental Defence, Friends of the Earth & International Rivers Network 2003. *Gambling with People's Lives. What the World Bank's New "High-Risk/High-Reward" Strategy Means for the Poor and the Environment.* Environmental Defence, Friends of the Earth & International Rivers Network. Washington, Amsterdam & Berkeley, p. 1, 27.

76. Bosshard, P. 2010. China Dams the World. *World Policy Journal*, Vol. 26, No. 4 (Winter 2009/2010): 43–51, p. 43.

77. Conca, K. 2006. *Governing water: contentious transnational politics and global institution building.* Mass.: MIT Press. Cambridge, p. 167–214.

78. Conca 2006: 185.

79. Majot, J. 2003. Comment in *World Rivers Review* Volume 18, Number 5: 1–16.

80. op. cit. Oweyegha–Afunaduula, F. C. 2000. Can huge dams solve our economic problems? Paper Presented at a seminar organised by NAPE and SBC to commemorate the 14th March 2000 International Day of Action Against Dams and for Rivers, Water and Life and held at the International Conference Centre, Kampala, Uganda, 14 March 2000. http://www.internationalrivers.org/resources/can-huge-dams-solve-our-economic-problems-1825 (accessed 27 January 2014)

81. Briscoe, J. 2010. Viewpoint – Overreach and Response: The Politics of the WCD and its Aftermath. *Water Alternatives* 3(2): 399–415.

82. WCD 2000. *Dams and Development. A New Framework for Decision-Making. The Report of the World Commission on Dams.* Eartscan Publications Ltd. London, p. xxix.

83. Oestigaard, T. 2012. Water Scarcity and Food Security along the Nile: Politics, population increase and climate change. *Current African Issues No. 49.* The Nordic Africa Institute. Uppsala.

84. WCD 2000: xxix–xxx.

85. UN World Water Development Report 2014. *Water and Energy. Volume 1.* Unesco. Paris, p. 99.

86. WCD 2000: xxxi.

87. WCD 2000: xxx.

88. WCD 2000: 321.

89. WCD 2000: 2.

90. WCD 2000: 321–322.

91. McCully, P. 2001. The Use of a Trilateral Network: An Activist's Perspective on the Formation of the World Commission on Dams. *American University International Law Review*, Vol. 16, Issue 6: 1453–1475, p. 1453.

92. McCully 2001: 1465.

93. WCD 2000: 281.

94. McCully 2001: 1454.

95. Conca 2006: 210.

96. McCully 2001: 1474–1475, my italics.

97. Briscoe 2010: 405.

98. Briscoe 2010: 405.

99. Briscoe 2010: 406.

100. Kader Asmal, chair of the World Commission on Dams, In Dams and Development: A New Framework for Decision-making, op. cit. Conca 2006: 167.

101. Briscoe 2010: 407.

102. Briscoe 2010: 407.

103. Briscoe 2010: 408.

104. Briscoe 2010: 409.

105. Briscoe 2010: 409–410.

106. Briscoe 2010: 410.

107. http://www.internationalrivers.org/resources/will-the-terminator-reduce-poverty-4096#note3 (accessed 8 April 2014)

108. Mutambi, B. 2012. Sustainable energy for all initiative. How prepared is Uganda? *Electricity Regulatory Authority.* Newsletter Issue 7, December 2012: 4–6, p. 5.

109. Mutyaba, V. 2012. Uganda at 50 and the Electricity sub-sector 58: what does the future hold? *Electricity Regulatory Authority.* Newsletter Issue 7, December 2012: 10–12, p. 12.

110. http://www.census.gov/population/international/data/idb/
region.php?N=%20Results%20&T=13&A=separate&RT=0&Y=2040&R=-1&C=UG

111. Borde, B. H. & Charafi, Y. 2013. An inconvenient truth. *Private Sector & Development. Proparco's Magazine.* No. 18/November 2013: 9–12.

112. Briscoe, J. 2011. Invited opinion interview: Two decades at the centre of world water policy. Interview with John Briscoe by the *Water Policy* Editor-in-Chief. *Water Policy* 13(2011): 147–160.

113. UN World Water Development Report 2014: 98–99.

114. Uganda Must Embrace Industrialisation – Museveni, 09–16 January 2013 http://www.africanexecutive.com/modules/magazine/articles.php?article=7015 (accessed 7 April 2014)

115. Oweyegha-Afunaduula and Isaac 2004.

116. *New Vision.* Tourism, the rich neglected goldmine. 20 September 2013. http://www.newvision.co.ug/mobile/Detail.aspx?NewsID=647430&CatID=3

117. WB IP 2008: 208.

118. WB IP 2008: 218.

119. African Development Bank 2009: iii.

120. Burnside 2006: 23–24.

121. Mallaby, S. 2004. NGOs: Fighting Poverty, Hurting the Poor. *Foreign Policy,* No. 114 (Sep. – Oct., 2004): 50–58, p. 52.

122. Independent Review Mechanism 2nd monitoring report on the implementation of findings of non-compliance and related actions to be undertaken by the ADB management on the Bujagali hydropower and interconnection projects 28 July, 2010; Independent Review Mechanism 3rd monitoring report on the implementation of findings of non-compliance and related actions to be undertaken by the AfDB management on the Bujagali hydropower and interconnection projects 06 June, 2011, and Independent Review Mechanism 4th[d] monitoring report on the implementation of findings of non-compliance and related actions to be undertaken by the AfDB management on the Bujagali hydropower and interconnection projects September 2012.

123. EIB-CM 2012: 29.

124. Mallaby 2004: 57.

125. Anderson, K. 2009. Review: What NGO Accountability Means: And Does Not Mean. NGO Accountability: politics, principles & Innovations, by Lisa Jordan & Peter van Tuijl (eds.). *The American Journal of International Law*, Vol. 103, No. 1: 170–178, p. 177.

126. Burnside 2006: 40.

127. Mutyaba 2012: 11.

128. Report No. IDA/R2008–0296: 3.

129. Mallaby 2004: 57.

130. Bosshard 2010: 43–51.

131. Briscoe 2010: 413.

132. http://www.monitor.co.ug/News/National/Bujagali+Dam+commissioning+/-/688334/1528064/-/12vonvaz/-/index.html (accessed 23 April 2014)

133. Conca 2006: 200.

134. Briscoe 2010: 411.

135. Oweyegha-Afunaduula, F.C. and Isaac, A. 2004. Environmental Hydropolitics of the Nile Basin's Bujagali Dam, Uganda: An Annotated Bibliography. Working to Protect River Nile, Fighting for Justice Occasional Paper No.7 NAPE/SBC-2004 February 6 2004.http://www.afuna.o-f.com/articles/ENVIRONMENTAL%20HYDROPOLITICS%20OF%20THE%20NILE%20BASIN%E2%80%99S%20BUJAGALI%20DAM,%20UGANDA.htm (accessed 30 January 2014)

136. Probe Alert September 2001.

137. Probe Alert September 2001.

138. Oweyegha-Afunaduula and Isaac 2004.

139. Briscoe 2010: 403.

140. Tangri, R. & Mwenda, A. M. 2013. *The Politics of Elite Corruption in Africa. Uganda in comparative African perspective*. Routledge. London, p. 100–101.

141. WB IP 2008: xxiv–xxv.

142. Number B-0130-UG. Indemnity Agreement (Partial Risk Guarantee for the Private Power Generation (Bujagali) Project) between International Development Association and Republic of Uganda. Dated July 18, 2007.

143. *Electricity Regulatory Authority. Developments and Investment Opportunities in renewable Energy Resources in Uganda*. June 2012.

144. Wakabi, M. 2013. Karuma power plant paves way for more stations. The East African 17 August 2013. http://www.theeastafrican.co.ke/business/Karuma-power-plant-paves-way-for-more-stations/-/2560/1957328/-/r8a3wv/-/index.html (accessed 11 April 2014)

145. China Helps Unlock Uganda's Economic Potential through Infrastructure Debt. *CRIENGLISH*. 11 May 2013. http://english.cri.cn/6826/2013/11/05/3441s796634.htm (accessed 11 April 2014)

146. Conca 2006: 208.

147. Oweyegha-Afunaduula and Isaac 2004.

148. Oweyegha-Afunaduula, F. C. 2003. Huge dams as corporate crime and terrorism. NAPE LOBBY, November 2003, 6th Edition Vol. 2 No. 4: p.21.

149. Oweyegha-Afunaduula and Isaac 2004.

150. Oweyegha-Afunaduula and Isaac 2004.

151. Report on Capacity Building Workshop for African Rivers Network Eastern Africa. Green Valley Hotel, Kampala, Uganda. 15th-17th July, 2010, p. 7–8. http://www.nape.or.ug/lib/ARN%20Capacity%20Building%20Workshop.pdf (accessed 20 January 2014)

152. WB IP 2008: xvii

153. WB IP 2008: 169.

154. WB IP 2008: 169.

155. WB IP 2008: 182–183.

156. WB IP 2008: 184.

157. WB IP 2008: 174.

158. WB IP 2008: 173, fn. 631.

159. WB IP 2008: 165–166.

160. WB IP 2008: 170.

161. Op. cit. WB IP 2008: 170.

162. WB IP 2008: 175.

163. Lacey, M. 2001. Traditional Spirits Block a $500 Million Dam Plan in Uganda. *The New York Times*, September 13. http://www.nytimes.com/2001/09/13/international/africa/13NILE.html (accessed 27 January 2014)

164. WB IP 2008: l.

165. WB IP 2008: liii.

166. WB IP 2008: 186–187.

167. Burnside 2006: 44.

168. Downing 2008, in WB IP 2008: 225.

169. Downing 2008: 225.

170. *The New Vision.* 27 November 2001. Uganda: Folks Eat AES Payoffs http://allafrica.com/stories/200111270649.html (accessed 21 February 2014)

171. Report No. IDA/R2008–0296. International Bank for Reconstruction and Development International Development Association. Management Report and Recommendation in Response to the Inspection Panel Investigation Report. Uganda. Private Power Generation (Bujagali) Project (IDA Guarantee No. B0130-UG). November 7, 2008, p. 22. http://siteresources.worldbank.org/EXTINSPECTIONPANEL/Resources/Management_Report_Nov_7_FINAL.pdf (accessed 10 April 2014)

172. WB IP 2008: 176.

173. WB IP 2008: 179.

174. WB IP 2008: 177.

175. WB IP 2008: 167.

176. WB IP 2008: 178.

177. WB IP 2008: 179.

178. Oweyegha-Afunaduula and Isaac 2004.

179. Lacey 2001.

180. WB IP 2008: 172.

181. WB IP 2008: 177.

182. WB IP 2008: 179–180.

183. Report No. IDA/R2008–0296: 25.

184. Mazige, J. 2003. Bujagali boss held over U.Shs. 1.4m debt. *The Monitor*, February 2003; p. 4. op. cit. Oweyegha-Afunaduula and Isaac 2004.

185. WB IP 2008: 180.

186. WB IP 2008: 174.

187. Downing 2008: 225.

188. Downing 2008: 224.

189. WB IP 2002. World Bank 2002. *Report No. 23988. The Inspection Panel. Investigation Report. UGANDA: Third Power Project (Credit No. 2268-UG), Forth Power Project (Credit No. 3545-UG) and Bujagali Hydropower Project (PRG No. B 003-UG), May 23, 2002.* World Bank. Washington, p. 95.

190. WB 2003. Accountability at the World Bank: The Inspection Panel 10 Years On. The World Bank. Washington, p. 83.

191. WB IP 2008: 176.

192. WB IP 2008: 179.

193. WB IP 2008: 186.

194. op. cit. WB IP 2008: 178, original italics.

195. IRM 2008. Independent Review Panel. Compliance Review Report on the Bujagali Hydropower and Interconnection Projects. June 20, 2008, p. 33.

196. WB IP 2008: 174.

197. http://www.ugandamissionunny.net/Parliament_address_February_2012_3_.pdf (accessed 7 April 2014)

198. Kitimbo, I. 2007. *The Monitor.* 9 September 2007. Uganda: Bujagali Spirits Relocated. http://allafrica.com/stories/200709100646.html (accessed 21 February 2014)

199. Ibid.

200. Kitimbo 2007.

201. IAU 2010. *Bujagali Hydropower Project (BHHP) Witness NGO Annual Report 2010. January-December 2010.* Inter Aid Uganda. Kampala, p. 27–28.

202. *European Investment Bank Complaints Mechanism (EIB-CM). Bujagali Hydroelectric Project, Jinja, Uganda. Complaint SG/E/2009/09.* Conclusion Report. 30 August 2012.

203. EIB-CM 2012: 24.

204. EIB-CM 2012: 104–106.

205. EIB-CM 2012: 107.

206. EIB-CM 2012: 104.

207. EIB-CM 2012: 107.

208. Mugabi, F. 2011. Bujagali spirits moved to new place. *New Vision.* 6 July. http://www.newvision.co.ug/D/8/17/759459 (accessed 10 April 2014)

209. Musinguzi, B. 2011. 'Budhagali' water spirits find a new resting place. The East African 24 July 2011. http://www.theeastafrican.co.ke/magazine/Budhagali+water+spirits+find+a+new+resting+place/-/434746/1206588/-/e6n3tvz/-/index.html (accessed 21 February 2014)

210. East Africa: 'Budhagali' Water Spirits Find a New Resting Place, 24 July 2011, http://allafrica.com/stories/201107261205.html (accessed 7 April 2014).

211. EIB-CM 2012: 106.

212. EIB-CM 2012: 107.

213. EIB-CM 2012: 106.

214. Gonza, R. K. et. al. 2002. *Traditional Religion and Clans Among the Busoga.* Vol. 1. Cultural Research Centre. Jinja, p. 152–153.

215. Gonza, R. K. et. al. 2001. *Reconciliation among the Busoga.* Cultural Research Centre. Jinja.

216. *Electricity Regulatory Authority. Developments and Investment Opportunities in renewable Energy Resources in Uganda.* June 2012, p. 13.

217. Heike Behrend pers. com.

218. Behrend, H. 1999. *Alice Lakwena & the Holy Spirits.* James Currey. Oxford, p. 30.

219. Behrend 1999: 25–26.

220. Behrend 1999: 62–63.

221. Behrend 1999: 134.

222. Behrend 1999: 134.

223. Behrend 1999: 91, 97.

224. Behrend 1999: 179, 185.

225. Weber, M. 1964. *The Sociology of Religion*. Beacon Press. Boston, p. 1.

226. Obeyesekere, G. 1984. *The Cult of the Goddess Pattini*. The University of Chicago Press. Chicago, p. 51, original emphasis.

CURRENT AFRICAN ISSUES PUBLISHED BY THE INSTITUTE

Recent issues in the series are available electronically
for download free of charge www.nai.uu.se

1981

1. *South Africa, the West and the Frontline States. Report from a Seminar.*
2. Maja Naur, *Social and Organisational Change in Libya.*
3. *Peasants and Agricultural Production in Africa. A Nordic Research Seminar. Follow-up Reports and Discussions.*

1985

4. Ray Bush & S. Kibble, *Destabilisation in Southern Africa, an Overview.*
5. Bertil Egerö, *Mozambique and the Southern African Struggle for Liberation.*

1986

6. Carol B.Thompson, *Regional Economic Polic under Crisis Condition. Southern African Development.*

1989

7. Inge Tvedten, *The War in Angola, Internal Conditions for Peace and Recovery.*
8. Patrick Wilmot, *Nigeria's Southern Africa Policy 1960–1988.*

1990

9. Jonathan Baker, *Perestroika for Ethiopia: In Search of the End of the Rainbow?*
10. Horace Campbell, *The Siege of Cuito Cuanavale.*

1991

11. Maria Bongartz, *The Civil War in Somalia. Its genesis and dynamics.*
12. Shadrack B.O. Gutto, *Human and People's Rights in Africa. Myths, Realities and Prospects.*
13. Said Chikhi, Algeria. *From Mass Rebellion to Workers' Protest.*
14. Bertil Odén, *Namibia's Economic Links to South Africa.*

1992

15. Cervenka Zdenek, *African National Congress Meets Eastern Europe. A Dialogue on Common Experiences.*

1993

16. Diallo Garba, *Mauritania–The Other Apartheid?*

1994

17. Zdenek Cervenka and Colin Legum, *Can National Dialogue Break the Power of Terror in Burundi?*
18. Erik Nordberg and Uno Winblad, *Urban Environmental Health and Hygiene in Sub-Saharan Africa.*

1996

19. Chris Dunton and Mai Palmberg, *Human Rights and Homosexuality in Southern Africa.*

1998

20. Georges Nzongola-Ntalaja, *From Zaire to the Democratic Republic of the Congo.*

1999

21. Filip Reyntjens, *Talking or Fighting? Political Evolution in Rwanda and Burundi, 1998–1999.*
22. Herbert Weiss, *War and Peace in the Democratic Republic of the Congo.*

2000

23. Filip Reyntjens, *Small States in an Unstable Region – Rwanda and Burundi, 1999–2000.*

2001

24. Filip Reyntjens, *Again at the Crossroads: Rwanda and Burundi, 2000–2001.*
25. Henning Melber, *The New African Initiative and the African Union. A Preliminary Assessment and Documentation.*

2003

26. Dahilon Yassin Mohamoda, *Nile Basin Cooperation. A Review of the Literature.*

2004

27. Henning Melber (ed.), *Media, Public Discourse and Political Contestation in Zimbabwe.*

28. Georges Nzongola-Ntalaja, *From Zaire to the Democratic Republic of the Congo.* (Second and Revised Edition)

2005

29. Henning Melber (ed.), *Trade, Development, Cooperation – What Future for Africa?*

30. Kaniye S.A. Ebeku, *The Succession of Faure Gnassingbe to the Togolese Presidency – An International Law Perspective.*

31. J.V. Lazarus, C. Christiansen, L. Rosendal Østergaard, L.A. Richey, Models for Life – Advancing antiretroviral therapy in sub-Saharan Africa.

2006

32. Charles Manga Fombad & Zein Kebonang, *AU, NEPAD and the APRM – Democratisation Efforts Explored.* (Ed. H. Melber.)

33. P.P. Leite, C. Olsson, M. Schöldtz, T. Shelley, P. Wrange, H. Corell and K. Scheele, *The Western Sahara Conflict – The Role of Natural Resources in Decolonization.* (Ed. Claes Olsson)

2007

34. Jassey, Katja and Stella Nyanzi, *How to Be a "Proper" Woman in the Times of HIV and AIDS.*

35. M. Lee, H. Melber, S. Naidu and I. Taylor, *China in Africa.* (Compiled by Henning Melber)

36. Nathaniel King, *Conflict as Integration. Youth Aspiration to Personhood in the Teleology of Sierra Leone's 'Senseless War'.*

2008

37. Aderanti Adepoju, *Migration in sub-Saharan Africa.*

38. Bo Malmberg, *Demography and the development potential of sub-Saharan Africa.*

39. Johan Holmberg, *Natural resources in sub-Saharan Africa: Assets and vulnerabilities.*

40. Arne Bigsten and Dick Durevall, *The African economy and its role in the world economy.*

41. Fantu Cheru, *Africa's development in the 21st century: Reshaping the research agenda.*

2009

42. Dan Kuwali, *Persuasive Prevention. Towards a Principle for Implementing Article 4(h) and R2P by the African Union.*

43. Daniel Volman, *China, India, Russia and the United States. The Scramble for African Oil and the Militarization of the Continent.*

2010

44. Mats Hårsmar, *Understanding Poverty in Africa? A Navigation through Disputed Concepts, Data and Terrains.*

2011

45. Sam Maghimbi, Razack B. Lokina and Mathew A. Senga, *The Agrarian Question in Tanzania? A State of the Art Paper.*

46. William Minter, *African Migration, Global Inequalities, and Human Rights. Connecting the Dots.*

47. Musa Abutudu and Dauda Garuba, *Natural Resource Governance and EITI Implementation in Nigeria.*

48. Ilda Lindell, *Transnational Activism Networks and Gendered Gatekeeping. Negotiating Gender in an African Association of Informal Workers.*

2012

49. Terje Oestigaard, *Water Scarcity and Food Security along the Nile. Politics, population increase and climate change.*

50. David Ross Olanya, *From Global Land Grabbing for Biofuels to Acquisitions of AfricanWater for Commercial Agriculture.*

2013

51. Gessesse Dessie, *Favouring a Demonised Plant. Khat and Ethiopian smallholder enterprise.*

52. Boima Tucker, *Musical Violence. Gangsta Rap and Politics in Sierra Leone.*

53. David Nilsson, *Sweden-Norway at the Berlin Conference 1884–85. History, national identity-making and Sweden's relations with Africa.*

54. Pamela K. Mbabazi, *The Oil Industry in Uganda; A Blessing in Disguise or an all Too Familiar Curse? Paper presented at the Claude Ake Memorial Lecture.*

55. Måns Fellesson & Paula Mählck, *Academics on the Move. Mobility and Institutional Change in the Swedish Development Support to Research Capacity Buildiing in Mozambique.*

56. Clementina Amankwaah. *Election-Related Violence: The Case of Ghana.*

57. Farida Mahgoub. *Current Status of Agriculture and Future Challenges in Sudan.*

58. Emy Lindberg. *Youth and the Labour Market in Liberia – on history, state structures and spheres of informalities.*

59. Marianna Wallin. *Resettled for Development. The Case of New Halfa Agricultural Scheme, Sudan.*

60. Joseph Watuleke. *The Role of Food Banks in Food Security in Uganda. The Case of the Hunger Project Food Bank, Mbale Epicentre.*

61. Victor A.O. Adetula. *African Conflicts, Development and Regional Organisations in the Post-Cold War International System. The Annual Claude Ake Memoral Lecture Uppsala, Sweden 30 January 2014.*

62. Terje Oestigaard. *Dammed Divinities. The Water Powers at Bujagali Falls, Uganda.*